Anaerobic Waste-Wastewater Treatment and Biogas Plants

Anaerobic Waste-Wastewater Treatment and Biogas Plants
A Practical Handbook

Joseph C. Akunna

CRC Press
Taylor & Francis Group
Boca Raton London New York

CRC Press is an imprint of the
Taylor & Francis Group, an **informa** business

CRC Press
Taylor & Francis Group
6000 Broken Sound Parkway NW, Suite 300
Boca Raton, FL 33487-2742

First issued in paperback 2020

ISBN 13: 978-0-367-73397-1 (pbk)
ISBN 13: 978-0-8153-4639-5 (hbk)

Library of Congress Cataloging-in-Publication Data

Names: Akunna, Joseph C. (Joseph Chukwuemeka), author.
Title: Anaerobic waste-wastewater treatment and biogas plants :
a practical handbook / Joseph C. Akunna.
Description: First edition. | Boca Raton, FL : CRC Press/Taylor & Francis Group,
[2018] | "A CRC title, part of the Taylor & Francis imprint, a member of the
Taylor & Francis Group, the academic division of T&F Informa plc." | Includes
bibliographical references and index.
Identifiers: LCCN 2018008211 (print) | LCCN 2018013732 (ebook) |
ISBN 9781351170512 (Adobe PDF) | ISBN 9781351170505 (ePub) |
ISBN 9781351170499 (Mobipocket) | ISBN 9780815346395
(hardback : acid-free paper) | ISBN 9781351170529 (ebook)
Subjects: LCSH: Sewage--Purification--Anaerobic treatment.
Classification: LCC TD756.45 (ebook) | LCC TD756.45 .A47 2018 (print) |
DDC 628.3/54--dc23
LC record available at https://lccn.loc.gov/2018008211

Visit the Taylor & Francis Web site at
http://www.taylorandfrancis.com

and the CRC Press Web site at
http://www.crcpress.com

Contents

Preface

The decision to write this book comes from discussions with colleagues, former students, waste and wastewater treatment managers, engineers, designers, and plant operators who believe that there is need for a book that brings all relevant information under the same "roof," and is written in a style that is easily understandable to both subject specialists and nonspecialists. Aspects of the subject are currently dispersed in different books, journals, manuals, etc. Some of these materials are often edited and with contributions from many authors. This makes it sometimes difficult to obtain a clear and consistence thread of information on key aspects of the technological applications. Furthermore, literature sources from technological providers are useful, but in most cases, they address their own products and services. This may present difficulties in making cost effective process and plant selection decisions.

This book addresses issues highlighted above by collating and presenting key aspects of anaerobic biotechnology through the perspective of a single author. The book also provides the requisite basic knowledge on biological treatment processes in general, as well as on appropriate process design and operational practices. The book's simple self-learning style encourages and enhances individual understanding of the essential aspects without the use of complex terminologies and equations. The key aspects considered in the book include *fundamental aspects of anaerobic and associated aerobic processes applied in waste and wastewater treatment; process design, operation, monitoring, and control; plant types, system configurations, and performance capabilities; co-digestion and feedstock pre- and post-treatment unit processes and operations; coproducts recovery, treatment, utilization, and disposal;* and *applications in warm and in developing countries.* This book also aims to guide specialists and nonspecialists on how anaerobic biotechnology can be part of solutions for the management of municipal and industrial solid, semisolid, and liquid residues in various climatic, environmental, and socioeconomic conditions.

The book is made of seven chapters, viz.:

Chapter 1 addresses the relevant fundamental aspects of biological treatment processes, anaerobic and associated non-anaerobic processes such as aerobic and anoxic processes, and the factors affecting their occurrences and efficiencies. The role of these non-anaerobic processes in the application of anaerobic biotechnology is also highlighted, and expanded in Chapters 4–6.

Chapters 2 and 3 address the applications of anaerobic treatment technology in wastewater, and solid and semisolid (such as biosolids) residues treatment, respectively. These chapters also highlight the key features common and different to both applications, particularly in terms of feedstock type and preparation, process design and operation, reactor type and system configuration, and performance monitoring.

Chapter 4 describes the common and emerging pretreatment processes and operations associated with solid and semisolid digestion. Their relative merits and demerits, as well as factors affecting their effectiveness are also covered.

Chapter 5 addresses the posttreatment requirements for gaseous, liquid, and semisolid coproducts of anaerobic treatment and options for beneficial use and disposal. Common and emerging posttreatment unit processes and operations are presented, as well as their merits and demerits.

Chapter 6 assesses the challenges to appropriate waste management in developing countries and explores the role of anaerobic technology in cost-effective and sustainable management practices and in general economic development. This chapter also highlights specific design, operational, and coproduct management factors in the application of the technology in developing countries and in warm climate regions.

Chapter 7 presents selected case studies from laboratory and pilot-scale studies.

Appendices A and B contain some worked examples on the application of anaerobic technology for the treatment of wastewater, and solid and semisolid residues, respectively.

This book has been made possible by the encouragement of my colleagues at the Urban Water Technology Centre, Abertay University and former students, to whom I say thanks. I also wish to thank my family for their encouragement and support particularly during the preparation of this book.

Joseph C. Akunna

Abbreviations

ABR	Anaerobic baffled reactor
AD	Anaerobic digestion
BMP	Biochemical methane potential (or Biomethane potential)
BOD	Biochemical oxygen demand
BOD$_5$	5 Day biochemical oxygen demand
BOD$_U$	Ultimate biochemical oxygen demand
CHP	Combined heat & power plant
C/N ratio	Carbon-nitrogen ratio
C/N/P ratio	Carbon-nitrogen-phosphorus ratio
C/N/P/S ratio	Carbon-nitrogen-phosphorus-sulfur ratio
COD	Chemical oxygen demand
CSTR	Continuously stirred-tank reactor
DF	Downflow filter
EGSB	Expanded granular sludge blanket
FBR	Fluidized bed reactor
F/M ratio	Food- microorganism ratio
FOS/TAC	Flüchtige organische säuren/Total anorganic carbon
GRABBR	Granular bed anaerobic baffled reactor
HRT	Hydraulic retention time
MBR	Membrane bioreactor
MC	Moisture content
MSW	Municipal solid waste
OFMSW	Organic fraction of municipal solid waste
OLR	Organic loading rate
RT	Retention time (or *same as* Hydraulic retention time)
SCOD	Soluble (or settled) COD
SS	Suspended solids
TA	Total alkalinity
TKN	Total Kjeldahl nitrogen
TOC	Total organic carbon
TS	Total solids
TSS	Total suspended solids
UASB	Upflow anaerobic sludge blanket reactor
UF	Upflow filter
US	Ultrasound
VFA	Volatile fatty acids
VS	Volatile solids
VSS	Volatile suspended solids

Author

Professor Joseph C. Akunna has a BEng (Hons) in civil engineering, MSc in hydraulics and water resources engineering, MSc and PhD in environmental engineering. He currently holds the Chair in Water & Environmental Engineering at Abertay University, Dundee, United Kingdom, where he is also the director of Postgraduate Environmental Engineering Education, and a founding member and co-director of the University's Urban Water Technology Centre. He has many years of research and teaching experiences in water and wastewater engineering, and in the treatment of municipal and industrial wastes and effluents. He is consulted widely by public and private sector organizations and has participated in and led many national and international research and development projects on the subject. Professor Akunna has published extensively and is a member of many academic and professional organizations in the subject area.

1

Biological Treatment Processes

1.1 Process Fundamentals

Biological processes involved in pollution control and bioenergy production are principally chemical reactions that are biologically mediated, and therefore can be otherwise referred to as biochemical processes. These biochemical reactions require external source of energy for initiation. In the case of biodegradable organics and other nutrients, the activation energy can be supplied by microorganisms that utilize these materials for food and energy. The sum total of processes by which living organisms assimilate and use food for subsistence, growth, and reproduction is called metabolism. Each type of organism has its own metabolic pathway, from specific reactants to specific end products. A generalized concept of metabolic pathways of importance in natural systems is shown in Figure 1.1.

Figure 1.1 shows that in biological processes, the substrates, and available electron donors will be transported into the microbial cell and all by-products, including extra microbial biomass and waste products, will be taken out of the cell by physical and chemical (physico-chemical) processes. Organic substrates can be in soluble or particulate form. The soluble and simpler monomers can diffuse through the cell walls into the cell, while the more complex fractions and particulate matter can physically adsorb on the cell surface where they are hydrolyzed to soluble and simpler monomers by extra-cellular enzymes produced by the microorganisms. These monomers are subsequently transported into the cell. Hence, the biodegradation of colloidal, particulate, and long-chain organic matter takes longer time than the biodegradation of simple sugars and volatile fatty acids.

The adsorption process plays an important role in the removal of both biodegradable and nonbiodegradable materials. Microorganisms and particulate organic matter in general, including those produced during biological oxidation, have a higher adsorption capacity than inert materials due to their relatively higher surface area. The combined physico-chemical and biological processes associated with adsorption of pollutants is referred to as *"biosorption."* Unless the adsorbed materials are biodegradable (i.e., they can be hydrolyzed and metabolized by microorganisms), their accumulation on

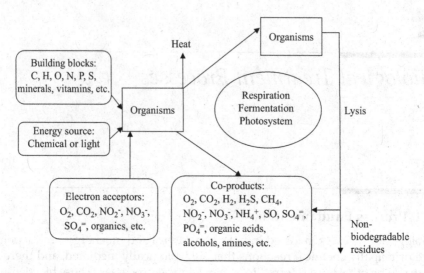

FIGURE 1.1
Generalized biochemical metabolic pathways.

microbial surfaces can eventually reduce the biosorption capacity of the biological treatment system. Changes in environmental and physico-chemical
conditions within the treatment system can however result in the release of
the adsorbed materials.

1.2 Anaerobic Processes

1.2.1 Process Description

Biological processes in the absence of molecular oxygen, where the electron
acceptors are carbon dioxide, organics, and sulfate (Figure 1.1) are generally referred to as *anaerobic processes*. These processes are similar to those
that occur naturally in stomachs of ruminant animals, marshes, organic
sediments from lakes and rivers, and sanitary landfills. The main gaseous
by-products are carbon dioxide (CO_2), methane (CH_4), and trace gases such as
hydrogen sulfide (H_2S), hydrogen (H_2), and a liquid or semiliquid by-product
known as digestate. The digestate consists of microorganisms, nutrients
(nitrogen, phosphorus, etc.), metals, undegraded organic mater and inert
materials.

Anaerobic processes typically occur in four steps, namely hydrolysis,
acidogenesis, acetogenesis, and methanogenesis, as shown in Figure 1.2.
Other processes that may also occur are nitrate and sulfate reductions, with
ammonia or nitrogen gas and hydrogen sulfide as by-products, respectively.

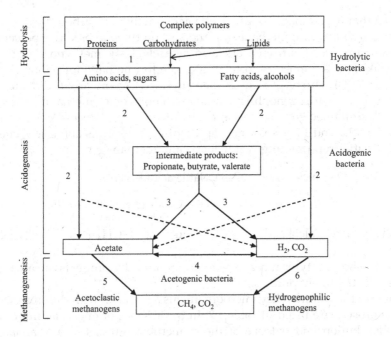

FIGURE 1.2
Simplified schematic diagram of different reactions involved in anaerobic digestion of complex organic matter. (1) Hydrolysis: complex polymers are hydrolyzed by extracellular enzymes to simpler soluble products. (2) Acidogenesis: fermentative or acidogenic bacteria convert simpler compounds to short chain fatty acids, alcohols, ammonia, hydrogen, sulfides, and carbon dioxide. (3) Acetogenesis: breakdown of short chain fatty acids to acetate, hydrogen, and carbon dioxide, which act as substrates for methanogenic bacteria. (4) Acetogenesis: reaction carried out by acetogenic bacteria. (5) Methanogenesis: about 70% of methane is produced by acetoclastic methanogens using acetate as substrate. (6) Methanogenesis: methane production by hydrogenophilic methanogens using carbon dioxide and hydrogen. (Adapted from Kasper and Wuhrmann, 1978; Gujer and Zehnder, 1983.)

Hydrolysis involves the breakdown of complex polymeric organic substrates such as proteins, carbohydrates, and lipids into smaller monomeric compounds such as amino acids, sugars, and fatty acids. This reaction is facilitated by extracellular specific enzymes produced by a consortium of varied hydrolytic bacteria. The monomers released during hydrolysis are converted by bacterial metabolism of acid-forming bacteria also known as fermentative bacteria into hydrogen or formate, carbon dioxide, pyruvate, ammonia, volatile fatty acids, lactic acid, and alcohols (Mata-Alvarez 2003). Carbon dioxide and hydrogen gases are also produced during carbohydrate catabolism. In acetogenesis, some of the compounds produced by acidogenesis are oxidized to carbon dioxide, hydrogen, and acetic acid (acetate) by the action of obligate hydrogen-producing acetogens. In addition, acetic acid is produced during the catabolism of bicarbonate and hydrogen by homoacetogenic bacteria. Methanogenesis leads to the formation of CH_4.

The methanogenic bacteria use acetic acid, methanol, carbon dioxide, and hydrogen to produce methane gas and carbon dioxide. Seventy percent of methane produced is from acetic acid by acetoclastic methanogenic bacteria, making it the most important substrate for methane formation (Mata-Alvarez 2003). Thirty percent is then produced from carbon dioxide and hydrogen by hydrogenophilic (or hydrogenotrophic) methanogenic bacteria. A simplified representation of the biochemical processes is given in Equations 1.1 and 1.2; the former representing hydrolysis and acidogenesis, and the latter representing acetogenesis and methanogenesis.

$$\text{Organics} + \text{nutrients} \rightarrow \text{Volatile acids} + \text{alcohols}$$

$$+ H_2 + CO_2 + H_2S + NH_3 + \text{cells} \qquad (1.1)$$

$$\text{Volatile acids} + \text{alcohols} + H_2 + CO_2 + \text{nutrients} \rightarrow CH_4 + CO_2 + \text{cells.} \quad (1.2)$$

Table 1.1 shows a typical composition of gaseous by-products of functional anaerobic treatment process.

Some of the common key microorganisms associated with different stages of the process are listed in Table 1.2. In the absence of microbial inhibition, the distribution and balance of these microbial groups in any anaerobic biological processes system will depend on the nature of the available the substrates and the environmental conditions (e.g., pH, temperature, potential redox, etc.).

Anaerobic microorganisms can be suspended or can take the form of biofilm. Biofilm systems utilize various types of organic and inorganic materials as support media. They are able to retain a greater amount of biomass, and are generally more effective than suspended growth systems in wastewater treatment and hence are referred to as *high-rate* systems.

TABLE 1.1

Typical Biogas Composition of a Functional Anaerobic Treatment Process

Component	Typical Range (% Volume)
Methane (CH_4)	50–75
Carbon dioxide (CO_2)	25–50
Nitrogen (N_2)	0–10
Hydrogen (H_2)	0.01–5
Oxygen (O_2)	0.1–2
Water vapor	0–10
Ammonia (NH_3)	Less than 1%
Hydrogen sulfide (H_2S)	0.01–3
Siloxanes	0–0.00002

Source: German Solar Energy Society (DGS) and Ecofys (2005), IEA Bioenergy (2006). Detailed analysis of biogas composition and factors affecting its beneficial use are discussed in Chapter 5.

TABLE 1.2

Anaerobic Digestion Stages and Typical Associated Microbial Species

Stage	Microbial Species
Hydrolysis	*Acetovibrio, Bacillus, Butyrivibrio, Clostridium, Eubacterium, Micrococcus, Lactabaccillius, Peptococcus, Proteus vulgaris, Ruminococcus, Staphylococcus, Streptococcus*, etc.
Acidogenesis	*Bacillus, Butyrivibrio, Clostridium, Eubacterium, Desulfobacter, Desulforomonas, Desulfovibrio, Lactabaccillius, Pelobacter, Pseudomonas, Sarcina, Staphylococcus, Selenomonas, Streptococcus, Veillonella*, etc.
Acetogenesis	*Methanobacillus omelionskii, Clostridium, Syntrophomonas buswelii, Syntrophomonas wolfei, Syntrophomonas wolinii*, etc.
Methanogenesis	Acetoclastic methanogens: *Methanosaeta, Methanosarcina*, etc. Hydrogenophilic methanogens: *Methanobacterium, Methanobrevibacter, Methanoplanus, Methanospirillium*, etc.

Source: Adapted from Wheatley (1991) and Stronach et al. (1986).

In anaerobic wastewater treatment, biomass produced during treatment must be separated from the treated wastewater before disposal or further treatment. The selection of suitable solid separation methods will depend on the process type and the desired treated effluent quality. A separate gravity sedimentation tank may be used; some of the separated solids can be returned to the anaerobic system, and excess is disposed of. Current research has seen the development of more efficient separation techniques. One of these is the use of anaerobic systems (or reactor) equipped with membrane separation facilities at the effluent outlet of the treatment system, e.g., the anaerobic membrane bioreactor (MBR) systems. These systems have the ability to retain a great amount of biomass within the reactor and also produce treated effluent with very low solids content. Detailed characterization of common anaerobic systems is covered in Chapters 2 and 3.

1.2.2 Biomass Production

Anaerobic processes generally result in lower cell production than the aerobic processes by a factor of 8–10 times. Consequently, lesser microbial nutrients are required in anaerobic processes as shown in Figure 1.3 (Speece 2008). In anaerobic systems, all the microbial groups described in Figure 1.2 work together; hence, the cell production reported by many authors are usually for the combined microbial populations, even though there are marked differences between the fast-growing acid-forming (acidogens) and the slow-growing methane-forming (methanogens) bacteria. The generalized combined yield in a single anaerobic treatment system is in the range of 0.05–0.10 volatile solids (VS)/g chemical oxygen demand (COD), and the average yield for acid-forming and methane-forming bacteria are in the ranges of 0.06–0.12 and 0.02–0.06 g VS/g COD, respectively. The yield varies with types of substrate and the amount of time the microorganisms spent in

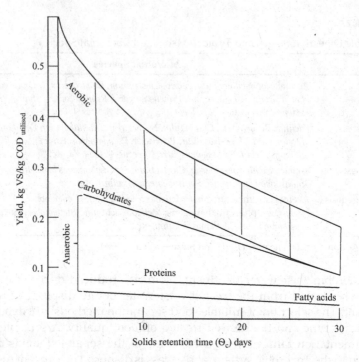

FIGURE 1.3
Biomass yield as a function of substrate type and solids retention time for aerobic and anaerobic biodegradation processes. (Adapted from Speece 2008.)

the biodegradation process as shown in Figure 1.3. This time is referred to as the solid retention time or SRT.

1.2.3 Factors Affecting Process Efficiency

1.2.3.1 Start-Up Inoculum

The performance and stability of anaerobic treatment processes depend on the quantity and quality of active methanogens present in the system. Active anaerobic microorganisms are often used to inoculate anaerobic systems during the start-up period in order to reduce the lag-phase in the development of a balanced microbial consortium. The microbial characteristics of an inoculum depend on the type of substrate and operational conditions of its source. Appropriate inoculum can be obtained from active anaerobic reactors, preferably those treating similar types of wastes or sewage sludge (biosolids). Anaerobic reactors are usually started by heavy seeding (at least 10% of the reactor volume or wastewater VS) or by maintaining the waste or wastewater pH in the range of 6.8 and 7.2 to encourage natural development of appropriate microbial populations, leading, respectively, to shorter and longer (up to 30 days) start-up time. Low inoculum-to-feed ratio may lead to the dominance of acidogens over methanogens and can result in low pH. When this occurs, recovery may be possible depending on the alkalinity of the system. Where alkalinity is low, chemical buffer may be added to the feed to avoid system failure. In *"dry"* solid anaerobic digestion (see Chapter 3), inoculum-to-feed ratios of greater than ten during start-up are recommended. In *batch* and *plug-flow* reactors or systems, fresh feed is usually pre-mixed (or pre-inoculated) with some of the digested residues.

1.2.3.2 Waste Organic Content and Biodegradability

Anaerobic treatment is most suitable for solid residues, slurries, and wastewaters with COD concentrations in the intermediate to high strength range, i.e., from 2,000 COD/L. Organic removal efficiencies tend to increase with increasing influent organic strength. However, up to 80%–90% COD removal is achievable in an efficiently operated system. *Posttreatment* employing aerobic processes may be necessary if further COD reduction is required. Posttreatment options are covered in Chapter 5. For low strength wastewaters (i.e., with COD <2,000 mg/L), aerobic treatment may be more appropriate. This is discussed further in Section 1.2 later.

The chemical composition of the waste or wastewater is one of the primary indicators of amenability of the organic constituents to biological treatment. Figure 1.4 shows the relative biodegradation rates and reaction times of various types of organic compounds. Biodegradability may be limited by the chemical structure of compounds such as lignin, cellulose, and hemicellulose, which are not readily amenable to enzymatic hydrolysis. These compounds may require prior treatment or *pretreatment*. Pretreatment techniques are discussed in Chapter 4.

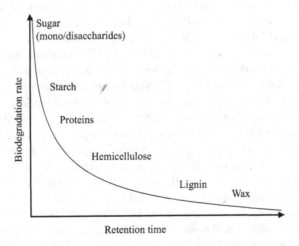

FIGURE 1.4
Relationship between rate of degradation and retention time for various types of organic compound. (Adapted from Eder and Schulz 2006.)

If a waste stream contains particulate matter, expressed in total solids (TS) (or total suspended solids [TSS]) or VS (or volatile suspended solids [VSS]), this must be hydrolyzed in the first stage of the process as shown in Figure 1.2. TS is a measure of all solids in the wastewater (and TSS measures only the *suspended* solids), while VS (or VSS) measures only the organic fraction (i.e., both biodegradable and nonbiodegradable) of TS (or TSS). Hydrolysis of particulate biodegradable organic matter is a relatively slow biological reaction for certain types of compounds, and may represent the process-limiting step in the treatment of high solids wastes. Such wastes usually require longer retention times to bring about high levels of treatment and biogas production. Conversely, if the organic constituents are primarily soluble in nature, high levels of biodegradation can be achieved at shorter retention times.

The *biochemical methane potential* or *biomethane potential* (BMP) assays are used for preliminary assessment of the anaerobic biodegradability and methane potential of organic residues. BMP assays and analysis are addressed in Chapter 4.

1.2.3.3 Nutrient Availability

The ideal feedstock composition ratio for carbon (C), nitrogen (N), phosphorus (P), and sulfur (S) (the C/N/P/S ratio) for hydrolysis and acidogenesis) is considered to be 500/15/5/3, and for methanogenesis, the ratio is theoretically assumed to be 600/15/5/3 (Weiland 2001). Some authors have expressed N and P requirements in proportion to COD as COD/N/P ratio of 580/7/1 (Hall 1992). C, N, and P are referred to as macronutrients. Sulfur and phosphorous requirements are very low compared to the other macronutrients; carbon, in particular, is naturally abundant in organic wastes. Therefore, the limiting nutrient for the anaerobic digestion process is considered mainly nitrogen. The carbon/nitrogen (C/N) ratio is used to measure nitrogen sufficiency of wastes to be treated by biological processes, with appropriate values ranging from 20 to 30 for anaerobic processes (Deublein and Steinhauser 2008; Polprasert 2007). Higher C/N ratios can lead to decreased bacterial growth due to nitrogen deficiency, while low ratios may result in ammonia toxicity on the microbial population. Wastes with high protein content have a relatively low C/N ratio and vice versa. Where there is a nitrogen deficiency, nutrient supplement may be needed and this is commonly achieved using urea, sewage sludge, or animal manure. Where a phosphorus deficiency exists, phosphorous can be added as phosphate salt or phosphoric acid. Careful combinations of wastes with extreme C/N ratios are usually preferred to achieve appropriate ratios.

Another important anaerobic microbial requirement is trace elements, or micronutrients, notably iron, cobalt, nickel, and zinc. These elements, when available in relatively small amounts, can stimulate methanogenic activities. The exact amount needed may vary for different wastes, and prior laboratory scale trials are needed before they are added to anaerobic treatment systems. Further discussions on micronutrients can be found in Chapter 4.

1.2.3.4 pH and Alkalinity

The stability of anaerobic treatment processes is highly dependent on pH. While the acidogenic bacteria are more tolerant to pH values below 6.0, optimum pH values for methanogenic bacteria lie between 7 and 8. Therefore, the pH range of 6.5–7.8 is suitable for the entire process. Acidic pH can occur in anaerobic treatment systems where a slower methanogenesis rate results in an accumulation of the volatile fatty acids. This situation is likely to occur where there is a sudden or excessive increase of organic matter added to the system. On the other hand, alkaline pH can result where the feed constitutes of high amounts of nitrogenous compounds, such as proteins. Nitrogenous compounds hydrolyze to produce ammonia, which causes alkaline pH. When pH values rise above 8.5, ammonia begins to exert a toxic effect on the methanogenic bacteria (Hartmann and Ahring 2006).

1.2.3.5 Temperature

Like all biological processes, anaerobic processes are affected by temperature. Figure 1.5 shows the relationship between temperature and the rate of anaerobic biodegradation. Anaerobic treatment systems can be operated at

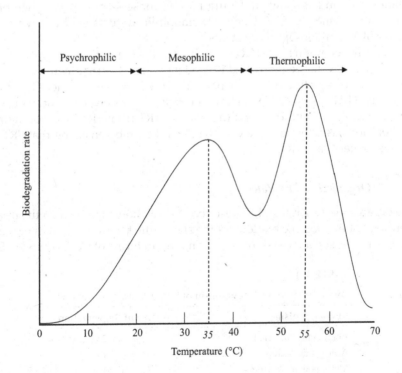

FIGURE 1.5
Temperature ranges for anaerobic treatment processes. (Adapted from Mata-Alvarez 2003.)

psychrophilic (<20°C), mesophilic (25°C–40°C), or thermophilic (45°C–60°C) temperature ranges, with the optimum temperatures for the mesophilic and thermophilic processes at about 37°C and 55°C, respectively (Abbasi et al. 2012; Mata-Alvarez 2003; Raposo et al. 2012). Psychrophilic temperatures are rarely employed due to the resulting low rate of biodegradation. The choice of either mesophilic or thermophilic operation is usually dependent on net economic gain that each can provide. In practice, most commercial plants operate at the mesophilic range.

1.2.3.6 Solids and Hydraulic Retention Times

The SRT refers to the average dwelling time of microorganisms within the system or reactor. SRT depends on microbial growth rate and on the rate of removal of excess microbial biomass from the treatment system. The former is affected by the nature of the constituent organic compounds as shown in Figures 1.3 and 1.4. Methanogenic bacteria have significantly slower growth rates than other microbial groups associated with anaerobic processes, as shown in Table 1.3. Consequently, the appropriate SRT is controlled by the need to reduce the rate of removal of the methanogenic microorganisms from the treatment system. Temperature also plays a vital role in the biodegradation rate, and as shown in Figure 1.5, and consequently on the microbial regeneration time and SRT. Hence, thermophilic systems tend to operate at shorter SRT than mesophilic systems.

Hydraulic retention time (HRT) is defined as the theoretical amount of time that the wastewater being treated is resident within the treatment system or reactor. For completely mixed suspended growth systems without biomass recycling, HRT is same as SRT. However, *high rate* wastewater treatment systems are designed and operated to separate HRT from SRT by encouraging greater biomass retention (see Chapter 2), and thereby ensuring that SRT is always longer than HRT.

1.2.3.7 Organic Loading Rate

The organic loading rate (OLR) describes the relationship between the treatment rate of the organic matter and the size of the treatment system or reactor. It is expressed as weight of organic matter in terms of COD or VS (or TS)

TABLE 1.3

Average Time of Regeneration of Some Microbial Groups

Microorganisms	Time of Regeneration
Acidogenic bacteria	Less than 36 hours
Acetogenic bacteria	80–90 hours
Methanogenic bacteria	15–16 days

Source: Adapted from Deublein and Steinhauser (2008).

added per volume of reactor per day. The higher the OLR a system can properly treat, the greater the cost-effectiveness of the system. High rate systems
generally cope with relatively high OLR.

1.2.3.8 Toxic Compounds

Anaerobic processes can be inhibited by substances contained in the waste,
which are toxic to microorganisms. Typical microbial inhibitors include
ammonia, sulfide, long-chain fatty acids, salts, heavy metals, phenolic compounds, and xenobiotics (Chen et al. 2008; Metcalf and Eddy 2014). The two
most common microbial inhibitors in anaerobic process are hydrogen sulfide
and ammonia.

1.2.3.8.1 Sulfide Toxicity

Sulfide is produced during the hydrolysis of sulfur compounds such as
sulfates, sulfite, and thiosulfate, which may be present in the waste, by the
activities of sulfate-reducing bacteria. In the absence of oxygen and nitrates,
sulfates serve as electron acceptor in the breakdown of organic materials.
Therefore, the sulfate-reducing bacteria compete with methanogenic bacteria for the available COD. Hence, the more the concentrations of sulfur
compounds in the waste, the greater the amount of COD required for its
hydrolysis and the lower the methane yield.

Hydrogen sulfide exists in aqueous solution in the forms of hydrogen
sulfide (H_2S), hydrogen sulfide ion (HS^-), and sulfide ion (S^{2-}), depending
on the pH, in accordance to the following equilibrium reactions shown in
Equations 1.3 and 1.4:

$$H_2S \leftrightarrow HS^- + H^+ \tag{1.3}$$

$$HS^- \leftrightarrow S^{2-} + H^-. \tag{1.4}$$

Figure 1.6 shows the effect of pH on H_2S, HS^-, and S^{2-} equilibrium in a 10^{-3}
molar solution. At pH of 8 and above, most of the sulfide produced remain in
solution, and below pH 8, it exists mainly in unionized gaseous form, H_2S. At
pH of 7, about 80% of the sulfides is present as H_2S. H_2S is considered more
toxic than the ionized forms, hence, pH is therefore an important parameter
in determining its level of toxicity. It is malodorous; toxic to humans, animals, and microorganisms; and corrosive to metals. Its concentrations in air
of above 20 mg/L should be avoided because of toxicity to humans. Although
up to 20 mg/L of sulfur compounds have been reported to be required for
cell synthesis in anaerobic treatment systems, its microbial inhibitory effect
is likely to occur when the waste COD/SO_4^{2-} ratio is less than 7.7 (Speece
2008). Anaerobic microbial activities can decrease by 50% or more at sulfide
concentrations of 50–250 mg/L. However, with adequate acclimatization,
this tolerance threshold can be surpassed in certain applications.

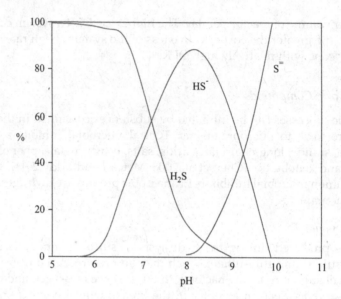

FIGURE 1.6
Effect of pH on hydrogen sulfide–sulfide equilibrium (10^{-3} molar solution, 32 mg H_2S/L. (Adapted from Sawyer et. al 1994.)

Sulfide inhibition can be minimized by the following measures (Pohland 1992):

- Dilution of the influent waste or wastewater
- Addition of iron salts into the treatment system to precipitate sulfide from solution
- Stripping the reactor liquid or scrubbing and recirculation of the reactor biogas
- Biological sulfide oxidation and sulfide recovery

Detailed analyses of sulfide-removal techniques in anaerobic systems are covered in Chapter 5.

1.2.3.8.2 *Ammonia Toxicity*

Ammonium (NH_4) is produced by the biological conversion of protein and urea under aerobic or anaerobic conditions. Ammonia (NH_3) is a gas and toxic to microbes, animals, and humans. NH_4 is related to the toxic un-ionized ammonia (also known as *free ammonia*) by the pH-dependent equilibrium shown in Equation 1.5.

$$NH_4^+ \leftrightarrow NH_3 + H^+. \tag{1.5}$$

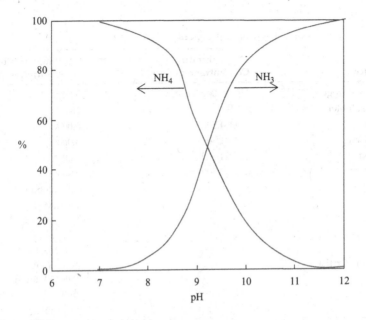

FIGURE 1.7
Effect of pH on ammonia–ammonium equilibrium. (Adapted from Metcalf and Eddy 2014.)

The proportion of NH_3, and hence the level of ammonia toxicity, rises with increase in pH as shown in Figure 1.7, and temperature. At pH of 7.5 and 30°C–35°C, 2%–4% of NH_4 will be in the form of NH_3 (Speece 2008). The toxicity threshold for free ammonia has been reported to be 100 mg NH_3-N/L although higher tolerance levels have also been observed due to microbial acclimatization. In terms of total ammonium, toxicity has been reported in the range of 1,500–3,000 mg NH_4-N/L at pH above 7.4, and over 3,000 mg NH_4-N/L. In general, the level of toxicity of a substance to microorganisms will depend on its nature, concentration, and the degree to which the process has become acclimated to the substance. As with most microorganisms, anaerobic bacteria can develop tolerance to a wide variety of inhibitors following adequate acclimatization. Levels of toxicity can also be reduced by decreasing the pH and temperature, and by diluting with water and/or incoming wastes or wastewaters.

1.2.3.8.3 Others

Light metal salts such as sodium, calcium, potassium, and magnesium are commonly found in anaerobic treatment systems either because of the breakdown of organic matter containing them or because they are constituents chemicals added to the system to bring about pH control. Traces of these light metals are required in anaerobic systems to stimulate bacterial growth and to ensure process optimization, but high levels can cause serious microbial inhibition. Accumulation of salts negatively affects microorganisms

TABLE 1.4

Inhibitors of Anaerobic Treatment Processes

Substance	Inhibiting Concentration (mg/L)	Strongly Inhibitory Concentration (mg/L)
Ammonia nitrogen	1,500–3,000 (at pH > 7.6)	3,000
Sulfide (soluble)	200	200
Calcium	2,500–4,500	8,000
Magnesium	1,000–1,500	3,000
Potassium	2,500–4,500	12,000
Sodium	3,500–5,500	8,000
Copper		0.5 (soluble)
Cadmium		150 mM/kg dry solids
Iron		1,710 mM/kg dry solids
Chromium, Cr(IV)		3 (soluble)
		200–250 (total)
Chromium, Cr(III)		2 (soluble)
		180–420 (total)
Nickel		30 (total)
Zinc		1 (soluble)

Source: EPA (1979), Parkin and Owen (1986).

because of the excessive increase of the osmotic pressure regulating the water flow across the cell membrane, which can lead to cell death (Ollivier et al. 1994). Inhibitory levels are dependent on the degree of acclimatization of the bacterial consortium and the synergistic effects resulting from the presence of other cations (Appels et al. 2008; Bashir and Martin 2004a; Bashir and Martin 2004b; Feijoo et al. 1995). Similarly, heavy metals such as chromium, cobalt, iron, zinc, or nickel can be found in relevant concentrations in some substrates. Their toxic effect is attributed to the disruption of enzyme function and structure (Chen et al. 2008).

Phenolic compounds can be grouped with long-chain fatty acids under the category of potentially inhibitory organic substances. They are inhibitory to microorganisms through their interaction with cell membrane inducing leakage of intracellular constituents (McDonnell 2007; McDonnell and Russell 1999). Other potentially toxic organics include halogenated benzenes, chlorophenols, and N-substituted aromatics. Table 1.4 lists some of the microbial inhibitory elements and compounds commonly encountered in anaerobic treatment systems, and their threshold concentrations.

1.2.3.9 Treatment Configuration: Single- and Multi-Stage Systems

Reactor configurations for single-stage and multi-stage systems are shown in Figure 1.8.

In single-stage systems, schematically shown in Figure 1.8a, all the processes described in Figure 1.2 take place within a single reactor. The advantage

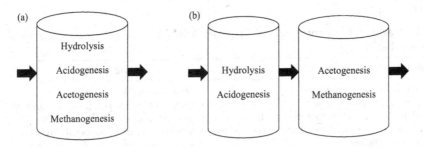

FIGURE 1.8
Single stage (a) and multi-stage (b) anaerobic treatment systems.

of a single reactor system is its relatively low capital and operational costs. A major drawback being that the system may not adequately cope with variations in substrate and environmental requirements, and kinetic properties of the different major microbial groups involved in the entire process, notably, the acidogenic and methanogen microorganisms. Providing and ensuring optimal conditions for each of these groups of microorganisms in a single reactor usually entails long SRT, HRT, and low OLR.

Multi-stage systems involve quasi separation of the key process stages in time and/or space. Space-based stage separation involves two or more reactors connected in series as shown in Figure 1.8b, or using reactor systems that can provide *plug flow* regime such as a compartmentalized reactor system shown in Figure 1.9. Examples of the latter include the anaerobic baffled reactor (ABR) (Barber and Stuckey 1999) or the granular-bed anaerobic baffled reactor (GRABBR) (Akunna and Clark 2000; Baloch and Akunna 2003; Shanmugam and Akunna 2008, 2010). Time-based stage separation is achieved in *batch* systems where feeding, reaction, solids-liquid separation and treated waste or wastewater withdrawal operations are carried out at selected time intervals. Stage or phase separation enables each of the different processes to take place at a place or time that is more suited to the microbial group involved. This method of operation can lead to greater process stability. The main drawback however, is an increased technical complexity and higher capital and operational costs, which may not always lead to increased economic gains.

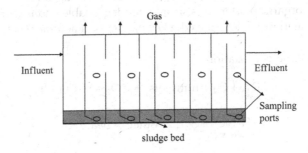

FIGURE 1.9
Anaerobic baffled reactor (ABR).

TABLE 1.5

Advantages and Disadvantages of Anaerobic Waste Treatment Processes Compared to Aerobic Treatment

Advantages	Disadvantages
• Low nutrient requirement	• Long start-up times
• Low sludge production	• Requires high temperatures for effective
• Methane production (potential fuel)	performance
• Treated effluent and digestate can be used as soil conditioner	• Sensitive to shock and variable organic load, changes in waste characteristics and temperature fluctuations
• No oxygen requirement, hence, low capital and operating costs	• Requires regular monitoring of input and by-products to ensure process stability
• Microorganisms can survive a long period of little or no feeding	• Requires skilled operational manpower
• Waste pasteurization can be achieved	

Source: Adapted from Hall (1992) and Malina (1992).

1.2.4 Applications, Benefits, and Drawbacks

Anaerobic technology is considered as a cost-effective solution for treating biodegradable organic residues, wastes, and wastewaters. In comparison with the alternative aerobic technology, the advantages and disadvantages of anaerobic processes are summarized in Table 1.5.

1.3 Aerobic Processes

1.3.1 Process Fundamentals

Aerobic biological treatment processes are carried out in the presence of molecular oxygen as an oxidizing agent or electron acceptor (see Figure 1.1). For the biodegradation of organic matter, or *carbonaceous oxidation*, the main end products are CO_2 and H_2O, and a liquid slurry or sludge consisting of microorganisms, nutrients (nitrogen, phosphorus, etc.), metals, inert materials, and undegraded organic matter consisting of biodegradable and nonbiodegraded materials. Equation 1.6 describes the key elements of aerobic reactions.

- Carbonaceous oxidation

$$\underset{\text{organic matter}}{COHNS} + O_2 + \text{nutrients} \rightarrow CO_2 + H_2O + NH_3$$

$$+ \underset{\text{new cells}}{C_5H_7NO_2} + SO_4^{2-} + \text{others} + \text{heat.} \tag{1.6}$$

As in anaerobic process, aerobic biodegradation of nitrogenous wastes produces ammonia. However, the ammonia can be further oxidized ultimately to nitrates (NO_3^-) in a sustained aerobic treatment in a process known as *nitrogenous oxidation or nitrification*. Nitrification is a two-stage process: formation of nitrite (NO_2^-) from ammonia (NH_3) by a group of bacterial known as *Nitrosomonas*, and followed by the oxidation of nitrite to nitrate by another group of microorganisms known as *Nitrobacter*. Both reactions are depicted, respectively, by Equations 1.7a and b. Equation 1.7c represents the overall process. The conversion of nitrite to nitrate is a swift process, and any accumulation of nitrite may be caused by oxygen deficiency, environmental conditions (e.g., pH, temperature), or microbial imbalance brought about most times by the presence of inhibitory compounds.

- Nitrification

Ammonia oxidation: $2NH_4^+ + O_2 \rightarrow 2NO_2^- + 4H^+ + 2H_2O + \text{new cells}$ (1.7a)

Nitrite oxidation: $2NO_2^- + 2O_2 \rightarrow NO_3^- + 2H^+ + H_2O + \text{new cells}$ (1.7b)

Overall: $NH_4^+ + 2O_2 \rightarrow NO_3^- + H_2O + 2H^+ + \text{new cells.}$ (1.7c)

Carbonaceous and nitrogenous oxidations are carried out by different microbial groups, *heterotrophs* and *autotrophs*, respectively. Autotrophic microorganisms can utilize carbon from inorganic sources such as carbon dioxide for their metabolic processes, while heterotrophic microorganisms utilize organic carbon for their synthesis. The conversion of carbon dioxide to cell carbon requires more energy than the heterotrophic process; hence, less energy is available for the production of new cells in autotrophic driven processes. Consequently, autotrophs have lower growth rates and cell mass yield than the heterotrophs. The heterotrophs usually outcompete the autotrophs for nutrients and oxygen, resulting in nitrification processes becoming more dominant only after most of the easily biodegradable organic compounds have been oxidized, as shown in Figure 1.10. During carbonaceous oxidation, organic nitrogen present is converted to ammonia, and any excess remaining after its use owing to its use in the production of new cells is converted to nitrates.

Where oxygen is still available after complete carbonaceous and nitrogenous oxidations, and without any new external substrate added to the system, the microorganisms will undergo *endogenous respiration* or *lysis* to release some of the nutrients used in the formation of the cell. The endogenous reaction is shown in Equation 1.8.

- Endogenous respiration

$$C_5H_7NO_2 + 5O_2 \rightarrow 5CO_2 + 2H_2O + NH_3.$$ (1.8)

If aeration is continued, nitrification will occur, as shown in Equation 1.9.

$$C_5H_7NO_2 + 7O_2 \rightarrow 5CO_2 + NO_3^- + 3H_2O + H^-. \tag{1.9}$$

Aerobic processes provide two types of applications in pollution control and waste/wastewater management: wastewater treatment and aerobic digestion.

1.3.2 Wastewater Treatment

As explained previously, carbonaceous and nitrogenous oxidations can result in the production of high amounts of new cells or *sludge* or *biosolids*. Where both carbonaceous and nitrogenous oxidations are required, the treatment duration or HRT is longer than if only carbonaceous oxidation is required, for the reasons given in Section 1.3.1. Up to 60% of the organic carbon constituent of the wastewater being treated can be used in the production of biosolids in aerobic processes.

The supply of oxygen and the management of resulting biosolids constitute major operational costs in aerobic treatment processes. It has been estimated that the energy requirement for oxygen supply can represent up to 65% of the total operational energy need where sophisticated, but highly effective artificial oxygen supply facilities are used in aerobic municipal wastewater treatment. Municipal wastewater is regarded as a medium strength wastewater because of its relatively low COD levels of less than 2,000 mg/L. Hence, the cost of oxygen supply and sludge management will

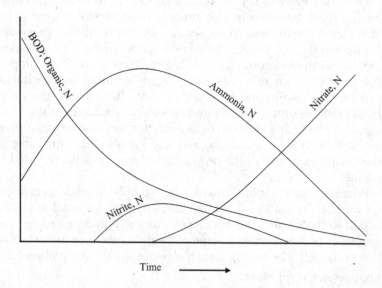

FIGURE 1.10
Fate of some of carbonaceous and nitrogenous compounds in aerobic treatment systems.

be much greater in the treatment of medium to high strength wastewaters. Oxygen can be introduced either artificially using mechanical energy intensive aeration systems or naturally by allowing atmospheric oxygen to diffuse into the wastewater treatment system. The latter is employed in ecological wastewater treatment systems such as lagoons and constructed wetlands. Systems based on natural oxygen transfer are commonly used to treat dilute wastewaters, and posttreatment (see Chapter 5) and their treatment efficiencies are directly proportional to the surface area of the treatment system. That is, the greater the surface-air contact area, the greater the potential for natural oxygen transfer and the greater the treatment efficiency. For medium to high strength wastes, wastewaters and slurries, artificial oxygen may be necessary to reduce space requirements and prevent uncontrolled anaerobic conditions from developing within the system.

Aerobic microorganisms can be either suspended or can take the form of biofilms on support media such as stones and plastic materials. Plastic materials have greater surface area/volume ratios than stones, and therefore systems will occupy less space than stone media systems.

As in anaerobic processes, the biosolids must be separated from the treated wastewater before discharging into receiving waterbodies. Appropriate solid separation methods will depend on the type of system and the desired treated effluent quality. Aerobic lagoons provide both biological oxidation and solids separation, while other processes may require separate settling or sedimentation tanks.

The use of membranes for solids separation is currently gaining acceptance with wastewater treatment plant designers. The addition of membranes in activated sludge systems removes the need for secondary sedimentation tanks and posttreatment that might be necessary for further solids reduction. Aerobic MBR systems are equipped with either internally immersed or external membrane separation units. The quality of treated effluent is usually high in terms of soluble and particulate organic content and may be suitable for a number of possible re-use applications. Early challenges in the use of the membrane technology lie in the high capital and operational costs of membranes. However, research within the last two decades has led to the development of more cost-effective MBR systems. More details on MBR systems can be found elsewhere.

1.3.3 Aerobic Digestion or Composting

Aerobic process applied for the treatment of solid and semi-solid wastes is referred to as aerobic digestion or *composting*. It is generally defined as the biological decomposition and stabilization of the organic constituents of solid wastes. As the organic material decomposes, the biological matter can heat up to temperatures between 50°C–70°C, and decreases as the amount of easily biodegradable organic matter decreases. Enteric pathogen contained in wastes usually be destroyed at this peak temperature range. Composting is carried out by a succession of mesophilic and thermophilic microorganisms and usually lead to pasteurization of organic residues.

Under prolonged aeration, nitrification of the ammonia by-product can also take place. Where the stoichiometric equation of the waste is known, Equations 1.10 and 1.11 (Tchobanoglous et al. 1993) can be used for quantitative estimation of composting oxygen demand and by-products.

- Carbonaceous oxidation

$$C_aH_bO_cN_d + 0.25(4a + b - 2c - 3d)O_2 \rightarrow aCO_2$$
$$+ 0.5(b - 3d)H_2O + dNH_3. \tag{1.10}$$

- Nitrification

$$NH_3 + 2O_2 \rightarrow H_2O + HNO_3. \tag{1.11}$$

Computation of the amount of oxygen required for aerobic digestion is illustrated in Appendix B. Further discussions on aerobic digestion are presented in Chapter 5.

Aerobic digestion is a relatively fast reaction and can be used as alternative to anaerobic digestion for the treatment of organic solid residues and biosolids. It can only lead to solid reduction where the amount of hydrolyzable solids in the untreated waste is greater than the amount of new aerobic microorganisms produced during the treatment process. It is usually preferred to anaerobic treatment where the waste solids are not readily amenable to anaerobic biodegradation, such as in the treatment of woody feedstock. Aerobic digestion can also be used as a pretreatment for anaerobic digestion, where its relatively high hydrolytic efficiency can be used to replace the much slower anaerobic hydrolytic stage for certain types of organic solids. Furthermore, the heat by-product of aerobic digestion can reduce anaerobic digestion heating requirements. Further discussions on the aerobic pretreatment in anaerobic digestion can be found in Chapter 4.

1.3.4 Aerobic versus Anaerobic Processes

Biosolids are the main by-products of aerobic treatment processes, while biogas and lower amounts of biosolids result from anaerobic treatment processes. Both processes require external energy input where high process efficiency is required: aerobic for oxygen supply and anaerobic for mesophilic or thermophilic operation. However, in anaerobic treatment, the energy requirement can be obtained from the produced biogas, where its production and recovery is economically viable. Moreover, the by-products of aerobic processes can be treated by anaerobic processes; just as the aerobic processes can be used for posttreatment of the liquid by-product of anaerobic processes.

For wastewaters, aerobic processes can ensure a better effluent quality than can be produced by only anaerobic processes in terms of carbonaceous and nitrogenous pollutants removal. In the treatment of high strength wastewaters therefore, it may be cheaper to firstly employ anaerobic treatment or *anaerobic pretreatment* and followed by aerobic treatment or aerobic pretreatment or *"aerobic polishing."* For low strength wastewaters, such as municipal wastewaters therefore, low biogas yield and mesophilic or thermophilic operation may result in anaerobic pretreatment being less cost-effective than direct application of aerobic treatment. However, in tropical countries with high ambient temperatures, anaerobic pretreatment of low strength wastewaters can offer a net positive energy gain due to the relatively low external energy required to bring the system to the more efficient mesophilic or thermophilic temperature ranges. Anaerobic pretreatment of municipal wastewater is currently in use in many countries, e.g., Brazil, Columbia, Ghana, etc.

In some instances, the objective of anaerobic pretreatment for low strength wastewaters may be simply to convert (i.e., bring about hydrolysis and acidification) complex organic molecules into simpler organic acids, and consequently reduce the oxygen (and energy) demand for a subsequent aerobic polishing, thereby encouraging the use of low energy and ecological systems for this purpose (see Chapter 6). The application of anaerobic pretreatment for this purpose may only require the anaerobic system to operate at ambient tropical temperature.

1.4 Anoxic Processes

Anoxic processes are biochemical reactions that occur in the absence of molecular oxygen and in presence of nitrogen oxides i.e., nitrate and/or nitrite. The nitrogen oxides serve as electron acceptors for the oxidation of organic or inorganic electron donors present, giving out gaseous nitrogen in the process. This process is called *denitrification* and is represented by Equation 1.12. Nitrates and nitrites can also be converted back to ammonia in a process known as *ammonification*.

- Denitrification

$$NO_3^- \rightarrow NO_2^- \rightarrow N_2O \rightarrow N_2. \tag{1.12}$$

This process is part of biological nitrogen removal employed in posttreatment of wastewaters to be discharged to certain receiving waterbodies. Biological nitrogen removal involves nitrification, followed by denitrification in an organic carbon-rich environment devoid of molecular oxygen. Since nitrification occurs in an environment that is deficient in biodegradable organic carbon, addition of external organic carbon sources is necessary to bring about denitrification. Using, for example, acetate as a source of external carbon, the denitrification reactions are represented as follows:

$$5CH_3COOH + 8NO_3^- + \rightarrow 4N_2 + 10CO_2 + 6H_2O + 8OH^-. \qquad (1.13)$$

Posttreatment options for anaerobically treated wastewater for carbon and nitrogen reduction are covered in Chapter 5.

2

Anaerobic Wastewater Treatment

2.1 Applications and Limitations

As explained in Chapter 1, anaerobic processes can play a significant role in the reduction of biodegradable organic matter from wastewaters. However, the treatment can achieve only about 80%–90% chemical oxygen demand (COD) removal; hence, post aerobic treatment may be necessary where higher levels of treatment are required. Some of the drivers for anaerobic treatment of wastewaters are as follows:

1. Pretreatment of high strength wastewater for energy recovery, followed by aerobic post treatment prior to discharge to the aquatic environment
2. Pretreatment of wastewaters to reduce oxygen demand and energy requirements in aerobic posttreatment prior to discharge to the aquatic environment
3. Sole treatment for stabilization and pasteurization of slurries prior to reuse for agriculture

Anaerobic processes are seldom used for domestic or municipal wastewater treatment except where the ambient temperature is of tropical nature and/or where its use will bring about a significant reduction in oxygen requirement for post-aerobic treatment. Anaerobic pretreatment can thus encourage the use of less energy intensive oxygen transfer systems, such as constructed wetland, for posttreatment. This is discussed in Chapter 5 in detail. Anaerobic processes are commonly used in the treatment of sludge (or biosolids) produced from aerobic processes for energy recovery in the form of biogas and pasteurization prior to their reuse for agriculture. Anaerobic treatment of solids and biosolids are covered in Chapter 3.

As shown in Figure 1.3, (Chapter 1), anaerobic microorganisms have generally lower growth rates than their aerobic counterparts. Therefore one of the key process design criteria is to separate the hydraulic retention time (HRT) from the solid retention time (SRT). HRT refers to the duration of treatment. It is controlled by the feeding or flow rate and its value takes into account the rate of biodegradation of the available substrates shown in Figure 1.4. The HRT is inversely proportional to the rate of biodegradation and directly proportional to the reactor size. The SRT is the amount of time microorganisms (or biosolids or biomass) produced during the biodegradation process stays in the system before being removed as excess biomass. It is controlled by the rate of removal of the produced biomass from the system. Controlled biomass removal at chosen intervals is referred to as *desludging* while uncontrolled excessive removal or *biomass washout* occurs via the treated effluent. The latter is facilitated by high hydraulic feeding rates (i.e., low HRT), inefficient effluent solid separation facilities, and undesirable microbial metabolic activities, such as *foaming*. To ensure effective operation, microorganisms must be present in sufficient quantity within the treatment system to ensure that the organic constituents or substrates are sufficiently broken down before the wastewater leaves the system. Low SRT and HRT values can result in low treatment efficiency and vice versa. *High-rate* reactors are generally designed and/or operated to ensure that both parameters can be separated, and where desired, to ensure that SRT is greater than HRT. Suspended growth systems equipped with biomass separation and recycling facilities, and fixed growth and membrane biological reactor (MBR) systems operating with treated effluent cycle are some of the methods that can be used to control HRT and SRT.

2.2 Wastewater Biodegradability

Anaerobic treatability of wastewater depends on the constituents of the wastewater as shown in Figure 1.4 (Chapter 1). While domestic, agricultural, food, and drink processing wastewater are generally amenable to anaerobic process, some industrial wastewater may contain compounds that are inhibitory to the microorganisms (Table 1.4, Chapter 1), and/or are poorly biodegradable and consequently require long HRT and SRT.

COD is commonly used to estimate the organic content of wastewaters, and measures both biodegradable and nonbiodegradable compounds. Biochemical oxygen demand (BOD) tests such as BOD_5 (5-day) and BOD_U (Ultimate) measure the readily biodegradable and total biodegradable organics, respectively. Unlike the COD, the BOD (i.e., BOD_5 or BOD_U) do not include reliable estimation of the amount of poorly biodegradable and nonbiodegradable organic matter. The BOD test is thus suited for the domestic wastewaters, which is composed mainly of readily biodegradable organics. It is, however, less reliable for assessing the suitability of industrial wastewaters for biological treatment for the following reasons:

1. Some industrial wastewaters may be deficient in some of the vital nutrients required for effective biodegradation.
2. BOD tests are adversely affected by the presence of (i) microbial inhibitory compounds and (ii) poorly biodegradable compounds thereby requiring longer time for microbial acclimatization beyond the limit of BOD test duration.

In practice, it is necessary to carry out an initial treatability test or the Biochemical methane potential or Biomethane potential (BMP) test. BMP tests are laboratory-based, and the results can provide vital information for both the design and operation of pilot or full-scale anaerobic treatment systems. Further detail on the BMP tests is provided in Chapter 4.

2.3 Wastewater Pretreatment

Wastewater pretreatment may be necessary to enhance the anaerobic treatment process. Common pretreatment techniques include sole or combinations of physical, chemical, or biological processes and operations as summarized below.

2.3.1 Flow Equalization

The purpose of flow equalization is to minimize and/or control fluctuations in wastewater characteristics passing through a treatment system. Anaerobic processes respond better to gradual changes to operating conditions. Abrupt and shock changes in wastewater characteristics can lead to process instability and poor performance. Flow equalization can prevent feed overload or underload, and ensure continuous operation even during periods of low or no wastewater production. For some industries, peak wastewater flows can occur only during daytime hours of weekdays, while low or zero production days occur in the nights, at weekends and during plant breakdown or routine shut down maintenance. Flow equalization ensures that the microorganisms in the treatment system are fed and kept alive at all times.

The size of flow equalization basins or tanks must be sufficient to accommodate the variability of wastewater streams and dilute the concentrated batches periodically produced. The tank must always be properly mixed to prevent short-circuiting, unwanted settling of solids, and uncontrolled fermentation, which can lead to odor nuisance, and health and safety issues. Mixing can be achieved by distribution of inlet flow and baffling, turbine mixing, mild diffused aeration with air or biogas. Where necessary and possible, flow equalization can be combined with nutrient and pH correction operations.

2.3.2 pH Correction

If the wastewater pH is already acidic, it may be harder for the pH of the reactor system to remain optimal for all microbial activities to proceed effectively, since more acidification is likely to occur during the biodegradation process. As stated in Chapter 1, the optimum pH for anaerobic treatment processes is in the range of 6.5–7.8. Where the wastewater is acidic, pH correction can be achieved by adding chemicals such as lime ($Ca(OH)_2$), sodium hydroxide (NaOH), or potassium hydroxide (KOH). Lime is a cheaper alternative, however, excessive precipitation of calcium carbonate ($CaCO_3$) can occur and this can cause scaling and accumulation of inorganic solids in the reactor. Where there is a consistent and reoccurring need for pH correction, it is preferable to use a mixture of cations for pH correction to avoid individual metal cation toxicity (see Table 1.4, Chapter 1). Recycling of treated effluent to dilute incoming wastewater and a reduction in feeding rate are alternative ways to bring about pH control without chemical addition.

pH control can also be carried out by adding alkaline materials to the wastewater where its alkalinity content is not sufficient to buffer acid produced during treatment. The assessment of the capability of anaerobic systems to adequately provide acid buffering capacity can be determined by the measure of its volatile fatty acids (VFA)/total alkalinity (TA) ratio, i.e., *VFA/TA ratio*. This ratio is also known as *FOS/TAC* (Flüchtige Organische Säuren/Total Anorganic Carbon) in German technical literature. The VFA is expressed in equivalent milligram of acetic acid per liter (mg acetic acid/L), and TA in equivalent milligram of calcium carbonate per liter (mg $CaCO_3$/L). Table 2.1 provides guidelines for the interpretation of the VFA/TA ratio of anaerobic treatment systems. The table indicates that if the ratio is below 0.3, it is assumed that there is low risk of rapid pH changes; hence, the anaerobic process is stable, and the system is functioning properly. For ratios less than 0.23,

TABLE 2.1

Guidelines for the Assessment of FOS/TAC Ratios

FOS/TAC Ratios	Interpretation	Remedial Measures
Greater than 0.6	Excessive organic overload	Stop feeding
0.5–0.6	High organic overload	Reduce feeding rate
0.4–0.5	High organic load	Monitor reactor performance indicators
0.3–0.4	Optimal organic load	Continue feeding at constant rate
0.2–0.3	Organic underload	Increase feeding rate gradually
Less than 0.2	Very low organic load underload	Increase feeding rate rapidly

Source: Adapted from Lossie and Pütz (2008).

the system is assumed to be underfed with no risk of pH-related instability, while ratios greater than 0.3 indicate a potential risk of low pH values and process instability. Detailed analysis of FOS/TAC, including analytical methods and its application in the assessment of the efficiencies anaerobic treatment plants can be found elsewhere, e.g., Rosato (2018), IEA Bioenergy (2013), and Lossie and Pütz (2008).

2.3.3 Nutrient Balance

Anaerobic treatment of microorganisms requires sufficient quantities of macronutrients, notably nitrogen and phosphorus, and micro or trace nutrients for growth as explained in Chapter 1. Nutrient deficiency can be overcome by adding external compounds or by co-treatment with other wastewaters containing the deficient nutrients. Process instability related to nutrient deficiency is commonly encountered in the treatment of some industrial wastewater, e.g., paper mill wastewater where addition of nitrogenous and micronutrients may be required. Prior compositional analysis of raw wastewater is therefore necessary to identify whether or not there is a need for nutrient adjustment.

2.3.4 Temperature Control

Although anaerobic processes can take place in psychrophilic, mesophilic, and thermophilic temperature ranges, mesophilic systems operating at optimum values of 35°C ± 2°C remain the most widely used for wastewater treatment. Lower temperature treatment systems are associated with low biodegradation rates and poor methane recovery, the latter worsened by the relatively high solubility of methane gas in water at low temperatures. Although high temperatures generally improve biodegradation rates, temperature fluctuations can cause process instability. Systems should normally be operated to avoid temperature changes of more than 1°C/day. Where the temperature of wastewater streams is variable due to the nature of processes and/or operations producing the wastewater, it is essential to provide a pretreatment in the form of cooling, heating, or mixing, preferably in equalization tanks, to reduce the temperature variability of wastewater fed to the treatment system.

2.3.5 Solids Reduction

Where the wastewater is constituted of significant concentrations of particulate biodegradable organic matter, the hydrolysis stage becomes the rate-determining step. Hydrolysis of particulate materials is a relatively slow biological reaction. Consequently, wastewaters with high solids content will require higher HRT and larger reactor size than those with low solids. Solids separation, and their disposal or separate treatment, can lead to reduced reactor size, resulting in reduction of the overall treatment cost. Many high-rate systems are not suitable for treating high solids wastewaters, therefore prior solids reduction may sometimes be necessary where these systems are to be used. The following methods can be used to reduce wastewater solids prior to anaerobic treatment:

1. *Sedimentation.* This is effective as a sole pretreatment where a significant proportion of the particles are settleable in sufficient amounts within a relatively short period of 2–4 h.

2. *Chemical precipitation.* This consists of coagulation, flocculation, followed by sedimentation. Chemical precipitation can achieve up to 80%–90% total suspended solids reduction within a relatively short period. Table 2.2 shows some of the characteristics and mode of actions of common coagulants. Certain coagulants are only effective at alkaline pH range; hence, pH correction before and after solids reduction may be required prior to anaerobic treatment. It is noteworthy that chemical precipitation can result in the precipitation of phosphorus, which is an important macronutrient in biological processes. Therefore, care must be taken to ensure that the pre-treated wastewater contains sufficient C/N/P ratio for biological treatment.

3. *Size reduction.* Particular size reduction increases the surface area of the particles available for biochemical actions. Examples of

TABLE 2.2

Chemical Precipitation Processes for Solids Reduction

Coagulant	Dosage range (mg/L)	pH	Comments
Lime	150–500	9.0–11.0	For solids reduction in wastewater with low alkalinity Reactions: $Ca(OH)_2 + Ca(HCO_3)_2 \rightarrow 2CaCO_{3\downarrow} + 2H_2O$ $MgCO_3 + Ca(OH)_2 \rightarrow Mg(OH)_2 + CaCO_{3\downarrow}$ At pH 11, $Mg(OH)_2$ and $CaCO_3$ are insoluble
Alum	75–250	4.5–7.0	For solids reduction in wastewater with high alkalinity Reactions: $Al_2(SO_4)_3 + 6H_2O \rightarrow 2Al(OH)_{3\downarrow} + 3H_2SO_4$
$FeCl_3$, $FeCl_2$ $FeSO_4.7H_2O$	35–150 70–200	4.0–7.0 4.0–7.0	For solids reduction in wastewater with high alkalinity. Presence of iron in the treated effluent Reactions: $FeCl_3 + 3H_2O \rightarrow Fe(OH)_{3\downarrow} + 3HCl$
Cationic polymers	2–5	No change	For solids reduction or as coagulant aid to metallic coagulants. No build-up of metallic ions in the effluent

Source: Adapted from Eckenfelder Jr. (1989).

size reduction treatment methods include physical (e.g., grinding, ultrasound, thermal, etc.), chemical (acid and alkaline treatments), and biological treatment (e.g., enzymatic hydrolysis). These operations and/or processes can be carried out within the wastewater or in a separate reactor where the removed solids are stored. For the latter, the treated solids can be returned to the wastewater streams for further treatment in anaerobic reactor. Detailed analysis of pretreatment techniques treatment are covered in Chapter 4.

2.3.6 Reduction of Toxic Compounds

Industrial wastewaters very often contain substances that are toxic to microorganisms. The toxicity of a substance depends on its nature, concentration, and the degree to which the biological process has become acclimated to it. As with most microorganisms, anaerobic bacteria can develop a tolerance to a wide variety of toxins following an adequate acclimatization.

Sulfide and ammonia toxicity are very commonly encountered in anaerobic treatment of food and drinks processing wastewaters. Sulfide toxicity can be overcome by recycling of treated wastewater to dilute incoming wastewater and/or by combining sulfate-rich and nonsulfate-rich wastewater streams. For example, in treating brewery wastewaters, combining wastewater streams from operations discharging excess yeast (which is rich in sulfate) with other less sulfate-containing wastewater streams can help reduce potential sulfide toxicity in treating excess yeast wastewater alone. Other possible remedial measures include: (i) addition of iron salts, such as ferric chloride directly inside the reactor to precipitate the produced sulfides and/ or (ii) addition of air or oxygen to the gas headspace of the reactor to oxidize and precipitate sulfide. Detailed description of sulfide reduction methods are presented in Chapter 5.

Ammonia toxicity can be reduced by mixing with ammonium deficient wastewater streams. Or posttreatment nitrification and denitrification followed by recycling of the treated effluent to dilute incoming wastewater. This option is further discussed in Chapter 5.

2.4 Process Variations

There are two types of anaerobic systems based on the physical occurrence of microorganisms, i.e., suspended and attached growth (or fixed film) systems. Some examples of these systems are shown in Table 2.3 and Figure 2.1. There are also *hybrid systems*, which are combinations of both systems. The performance of attached growth or fixed film systems depends on the physical and chemical properties of the carrier media (e.g., porosity, surface area,

TABLE 2.3

Types of Anaerobic Wastewater Treatment Systems

Conventional Systems: Suspended Growth	High-Rate Systems: Suspended Growth	High-Rate Systems: Attached Growth
a. Septic tanks	a. Multi-stage reactors	a. Fixed bed reactors
b. Continuously stirred tank reactor (CSTR)	b. Baffled reactors	b. Expanded bed or fluidized bed reactors
c. Sewage sludge digesters	c. Upflow sludge blanket reactors	c. Rotating bed reactors
d. Anaerobic covered lagoon and ponds	d. Expanded granular bed reactors	
	e. Reactors with internal recirculation	
	f. Sequencing batch reactors	
	g. Anaerobic contact process	
	h. Plug flow reactors with recirculation	
	i. Membrane bioreactors (MBR)	

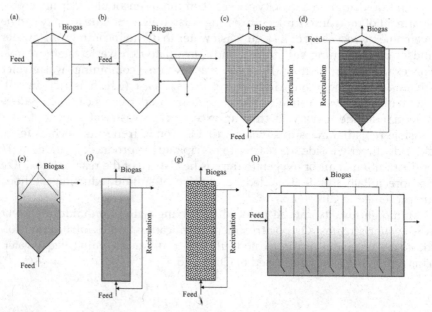

FIGURE 2.1

Types of anaerobic systems (a) Continuous stirred tank reactor (CSTR) without biomass recirculation (or conventional system), (b) contact process, (c) upflow filter, (d) downflow filter, (e) upflow anaerobic sludge blanket (UASB), (f) expanded granular sludge blanket (EGSB), (g) fluidised bed, and (h) baffled reactor (ABR) (adapted from Stamatelatou et al. 2014).

size, density, chemical composition, etc.), methods of mixing (e.g., fluidization, upflow or downflow velocities), hydraulic control features, scouring and clog prevention facilities utilized, etc. Each of these can be utilized as sole or combined with the same or different systems depending on costs and the desired process outcome. Therefore, the design and operational criteria of each system depends on many factors such as the type of reactors, system configuration, wastewater characteristics, treatment temperature, desired treated effluent quality, land availability, manpower requirements, etc. Some system design and performance criteria, particularly for high-rate systems, are often based on empirical formulae obtained from laboratory and field trials. Detailed design of some of high-rate systems can be found elsewhere (e.g., Malina and Pohland 1992; Metcalf and Eddy 2014; von Sperling and de Lemos Chernicharo 2005) and system manufacturers' brochures. Table 2.4 summarizes some the characteristics of typical systems.

TABLE 2.4

Summary of Key Characteristics of Some Anaerobic Treatment Systems

Technology	Characteristics	Possible Challenges
Conventional system	a. Low energy consumption b. Suitable for sewage sludge and high solids wastewater treatment c. SRT is generally equal to HRT	a. Biomass washout b. Low OLR/High HRT c. Low treatment efficiency d. High space requirement
Anaerobic contact process	a. CSTR with biomass recirculation b. SRT can be separated from HRT c. Suitable for low solids and for low to medium strength wastewaters	a. No phase separation b. High space requirement than other high-rate systems c. Not suitable for high OLR d. Poor settling in the settling tank due to *"rising sludge"* caused by gas bubbles e. Settling tank must be covered to prevent odor nuisance
Upflow anaerobic sludge blanket (UASB)	a. Formation/use of active and settleable biomass in the form of granules b. Suitable for low solids high strength wastewaters c. SRT is greater than HRT d. Low energy requirement e. Suitable for low to medium OLR	a. No phase separation b. Performance dependent on granule formation with wastewater being treated c. Suitable inoculum is required for rapid start-up, otherwise start-up may take up to 6 months d. Sludge blanket may be difficult to maintain during operation
Expanded granular sludge blanket (EGSB)	a. Modifications of UASB with higher upflow velocities to enhance hydraulic properties b. Suitable for medium to high OLR c. SRT is greater than HRT d. Good process control e. Low space requirement f. Suitable for low strength wastewaters	a. No phase separation b. Poor process stability at high OLR and low HRT c. Less flexible operation d. Not suitable for high solids wastewaters e. High energy requirement

(Continued)

TABLE 2.4 (*Continued*)

Summary of Key Characteristics of Some Anaerobic Treatment Systems

Technology	Characteristics	Possible Challenges
Fluidized bed reactor (FBR)	a. Development of anaerobic biomass on inert particles, and fluidization as in EGSB b. Suitable for medium to high OLR c. SRT is greater than HRT d. Good process control e. Low space requirement f. Suitable for low strength and low solids wastewaters	a. Difficulties in maintain optimum mixing and fluidizing velocity and conditions without biomass stripping or washout b. General difficulties in scaling up from pilot to effective full-scale operating conditions c. Difficult to start-up (Also see EGSB)
Anaerobic baffled reactor (ABR)	a. Plug flow regime b. Phase separation c. SRT can be separated from HRT via sludge recirculation d. Mechanical mixing not required e. Suitable for low strength and low solids wastewaters f. Possible selective recovery of high methane content biogas	a. Difficulties with variable system hydraulics due to sludge accumulation and biomass growth b. Limited full scale experience
Anaerobic upflow filter (UF)/ downflow filter (DF)	a. Anaerobic biomass attached on fixed inert media and quasi plug flow b. Performance is dependent on the type of media c. attached and suspended biomass, while DF involves mainly attached biomass d. Suitable for medium to high OLR e. Extremely high SRT:HRT ratio f. Good process control g. Low space requirement h. Suitable for low strength and low solids wastewaters	a. No phase separation b. Not suitable for high solids wastewaters c. Mixing and short circuiting problems d. High energy requirement

2.5 System Configuration

In one reactor system, all biological processes take place within a space. Mixing ensures that the concentration of all determinants in the aqueous solution remain the same in all parts of the reactor. There is thus a greater tolerance of accidental organic shock loads and of toxic compounds through dilution by the entire volume of wastewater in the reactor. However, since all processes occur within the same reactor, the need to balance the activities of hydrolytic, acidogenic, and methanogenic microorganisms can result in operating the system at low organic loading rate (OLR). This may lead to a need for large reactor sizes. Additionally, the effects of inhibitory compounds (such as ammonia, hydrogen sulfide) and unsuitable operating conditions (e.g., variable temperatures, out of range pH, etc.), are more likely to affect the acidogen–methanogen balance in favor of the acidogenesis, and this may lead to volatile fatty acids (VFA) build-up and process instability.

Separation of acidogenesis and methanogenesis increases process stability. This can be achieved using multiple reactors connected in series or baffled reactor systems that encourage plug flow within the reactor. These systems can also enable the recovery of biogas with higher energetic value since biogas from the acidogenic (rich in CO_2 and H_2S) and methanogenic zones (up to 86% CH_4) can be collected separately (Shanmugam and Akunna 2008). These systems, however, require higher capital and operation costs.

2.6 Process Design and Operational Control

2.6.1 Hydraulic Retention Time (HRT)

Hydraulic retention time (HRT), or simply *retention time* is the average time that the liquid stays in the treatment reactor before being discharged. It is also known as the *treatment time*, and can be calculated using the following equation:

$$\theta = \frac{V}{Q}, \tag{2.1}$$

where
  θ = Hydraulic retention time, h
  V = Reactor volume, m^3
  Q = Average wastewater flow rate, m^3/h

HRT can vary from 4 to 48 h, depending on the wastewater characteristics. Equation 2.1 shows that HRT is directly proportional to reactor size. The higher the concentration of organic matter in the wastewater, the longer the necessary treatment time. High wastewater solids require usually longer treatment time due to the extra time needed for solids hydrolysis. Very low HRT can lead to high biomass washout in suspended growth systems that are not equipped with suitable biomass retention or recovery recirculation facilities.

2.6.2 Solids Retention Time (SRT)

SRT or *sludge age* provides an estimate of the average time microorganisms or *biomass* produced during the biodegradation process stay within the system before being removed as waste or excess biomass. The use of SRT for process design and operation design is only applicable in suspended growth systems, where it is relatively easy to estimate the amount of biomass in the reactor. It is controlled by the rate of removal of excess biomass or *desludging* from the system as shown in the following equation:

$$\theta_c = \frac{VX}{(Q - Q_w)X_e + Q_w X_w},$$ (2.2)

where
θ_c = Solids retention time, day
V = Reactor volume, m^3
Q = Average wastewater flow rate, m^3/day
Q_w = Average waste biomass flow rate or *desludging rate*, m^3/day
X = Average concentration of biomass in the reactor, kg VS/m^3
X_e = Average concentration of biomass in the treated effluent, kg VS/m^3
X_w = Average concentration of biomass in waste or excess biomass stream, kg VS/m^3

Where $Q_w \lll Q$, and $X_e \lll X$, Equation 2.2 approximates to the following equation:

$$\theta_c = \frac{VX}{Q_w X_w}.$$ (2.3)

The preceding equation shows that in absence of excessive hydraulic load or very low HRT that can cause unwelcome biomass washout, excess biomass withdrawal rate is the single most important tool for controlling the SRT. The choice of biomass withdrawal rate is dictated by organic loading and biodegradation rates. High-rate reactors are generally designed and operated in a manner that will ensure high biomass retention, i.e., SRT >>> HRT. This is normally achieved by recirculation of biosolids separated from treated effluent or by the use of attached growth systems.

2.6.3 Hydraulic Loading Rate (HLR)

Hydraulic loading rate (HLR) measures the amount of liquid applied per unit area of the reactor, as expressed in Equation 2.4. It is commonly used in the design of fixed bed reactors.

$$HLR = \frac{Q}{A},$$ (2.4)

where

HLR = Hydraulic loading rate, $m^3/m^2 \cdot day$
Q = Average wastewater flow rate, m^3/day
A = Surface area of the packing medium, m^2

The higher the HLR, the lower the HRT. Optimal values for HLR vary with wastewater characteristics and types of support media. Plastic media have higher surface area to volume ratio than stone media. Hence, plastic media anaerobic reactors are usually operated at higher HLR than stone media reactors.

2.6.4 Organic Loading Rate (OLR)

OLR or *Volumetric Organic Load* represents the amount of biodegradable organic matter, expressed in terms of COD or BOD, applied daily per unit volume of the reactor, as expressed in the following equation:

$$OLR = \frac{QS_o}{V}. \qquad (2.5)$$

where

OLR = Organic loading rate, kg $COD/m^3 \cdot day$ (or kg $BOD/m^3 \cdot day$)
S_o = Influent BOD or biodegradable COD in wastewater, mg/L
V = Reactor volume, m^3
Q = Average wastewater flow rate, m^3/day

Suitable OLR values depend on many factors, notably wastewater characteristics, operating temperature, and the level of microbial activity within the reactor. OLR values are usually kept low during start-up and gradually increased as the reactor stability increases evidenced by optimum pH range and low build-up of VFA.

2.6.5 Food/Microorganism Ratio

The Food/Microorganism (F/M) ratio or *sludge loading rate* (SLR) represents the amount of biodegradable organic matter, expressed in terms of COD, applied daily per unit biomass present in the reactor, as expressed in Equation 2.6. It is only used in suspended growth systems where the amount of biomass can be more accurately estimated. The total volatile solids (VS) concentration in the reactor is assumed as a measure of the biomass content.

$$F/M\,ratio = \frac{QS_o}{VX}, \qquad (2.6)$$

where
  F/M = Food/Microorganism ratio, kg COD/kg VS·day (or 1/day)
  S_o = Influent biodegradable COD in wastewater, kg/m^3
  Q = Average wastewater flow rate, m^3/day
  V = Volume of reactor, m^3
  X = Average concentration of microorganisms present in the reactor, kg VS/m^3

In anaerobic systems, the methods used for the determination of the parameter X do not distinguish between the fast-growing acidogens and the slow-growing and rate-limiting methanogens. Hence, the F/M ratio is seldom used in the design and operation of anaerobic systems. It is, however, an important design and operational control parameter in aerobic treatment and posttreatment processes (see Chapter 6), where there is less distinction between the activities of the participating microorganisms.

2.6.6 Specific Biogas Yield

The specific biogas yield measures the maximum biogas production capability of a given amount of organic compound. It is estimated using the following equation:

$$Y_{biogas} = \frac{Q_{biogas}}{Q(S_o - S_e)},\qquad(2.7)$$

where
  Y_{biogas} = Specific biogas yield, m^3 biogas/COD$_{removed}$
  Q_{biogas} = Biogas production rate, m^3/day
  Q = Average wastewater flow rate, m^3/day
  S_o = Influent COD in wastewater, kg/m^3
  S_e = Effluent COD in wastewater, kg/m^3

The theoretical value of Y_{biogas} is a constant and stoichiometrically equals to 0.5 m^3/COD$_{removed}$, comprising 0.35 (or 70%) and 0.15 (or 30%) for methane and carbon dioxide, respectively. A comparison of actual Y_{biogas} value with the theoretical value for various types of wastewaters is important in understanding system performance and assessing the accuracy of monitoring devices.

2.6.7 Specific Biogas Production Rate (BPR)

BPR is used to compare the rates of biogas production by different anaerobic treatment systems. It is given by the following equation:

$$BPR = \frac{Q_{biogas}}{V},\qquad(2.8)$$

where
BPR = Specific biogas production yield, m^3 biogas/m$^3 \cdot$ day·
Q_{biogas} = Biogas production rate, m^3/day
V = Reactor volume, m^3

2.6.8 Treatment Efficiency

Treatment efficiency is a measure of the proportion of the target determinant removed or transformed in the treatment system. In anaerobic wastewater treatment, the treatment efficiency is the amount of *settled* COD removed in the system, and expressed in percentage as shown the following equation:

$$\% \text{COD removal} = \frac{S_o - S_e}{S_o} \times 100, \tag{2.9}$$

where
S_o = Influent settled wastewater COD, mg/L
S_e = Effluent settled wastewater COD, mg/L

2.6.9 Temperature

As discussed in Chapter 1, anaerobic treatment systems can be operated in the following temperatures:

a. Psychrophilic: 5°C–20°C

b. Mesophilic: 20°C–45°C (Optimum: 35°C ± 2°C)

c. Thermophilic: 45°C–70°C (Optimum: 55°C ± 2°C)

The values of most of the aforementioned design parameters are also dependent on the operating temperatures. For example, thermophilic and mesophilic systems have higher tolerance for lower HRT and higher OLR than psychrophilic systems.

2.7 Performance and Process Monitoring Indicators

A generalized performance template for anaerobic wastewater treatment is shown in Table 2.5.

Several parameters have been used as indicators of process imbalance in anaerobic systems. Some of the most commonly used indicators are listed in Table 2.6. Most of the parameters are inter-related, indicating that no parameter alone can provide a complete assessment of a treatment system.

TABLE 2.5

Typical Anaerobic Treatment Performance Levels

Treatment Parameter	Typical Value
BOD removal, %	80%–90%
COD removal, mg/L	$1.5 \times BOD_{removed}$
Biogas production	$0.5\,m^3/kg\;COD_{removed}$
Methane production	$0.35\,m^3/kg\;COD_{removed}$
Sludge production	$0.05–0.10\,kg\;VS/kg\;COD_{removed}$

Source: Pohland (1992).

TABLE 2.6

Indicators for Process Imbalance in Anaerobic Digestion

Indicator	Principle
Biogas production	Changes in specific gas production can be as a result of changes in feedstock characteristics, mainly the nature and biodegradability of the organic constituents (i.e., VS and COD)
Biogas composition	Changes in the biogas CH_4/CO_2 ratios are signs of process instability. Higher CO_2 content may be indicative of organic overload or organic shockload and/or inhibition of methanogenesis, caused by some or all of the following: high levels of VFA accumulation, ammonia, sulfide, or other inhibitors and changes in temperature. Hydrogen (H_2) in the biogas is usually low, and where detected, is also a sign of process instability commonly associated with high VFA accumulation
pH	Changes depend on VFA and ammonium concentrations. Drop in pH can be due to VFA accumulation, and/or drop in alkalinity. High CO_2 biogas content can be related to low pH
Alkalinity (TA)	Detects changes in buffer capacity, and can affect pH levels
VFA	Accumulation indicates process instability, which can be caused by one or more of the following: lower methanogenic activities and/or higher acidogenic than methanogenic activities, lower acetogenic activities for some individual VFAs (commonly propionic acid), lower alkalinity of medium, temperature changes, organic overload and/or shockload, etc.
Individual VFA	Accumulation of individual VFA
VFA/TA (or FOS/TAC) ratio	Values of about 0.3 indicate process stability, and greater values are indicative of (looming) process instability (see Table 2.1)
COD or VS content	Changes in biodegradation rate
Temperature	A variation of 2°C–3°C can lead to fundamental changes in the microbial dynamics of the system. Different temperature ranges are associated with different microbial populations, particularly for methanogenesis. Excessive VFA accumulation and high biogas CO_2 content can also be caused by temperature changes

While biogas production, composition, and reactor temperature are relatively simple to measure, the others are more time consuming and expensive to quantify. The simpler measurements can be carried out more frequently, while others less routinely and when the simpler measurements suggest looming instability.

2.8 Foaming and Control

Foaming is an undesirable occurrence in anaerobic wastewater treatment systems which can cause both physical effects e.g., overflow of effluents to surrounding areas, blockage of biogas pipework, interference with monitoring and process control devices, high solids in the effluent pipes, etc. and biological effects with overall impact of reduced biological efficiency. Overall, foaming can lead to high operational costs and general losses in earnings.

There has been a lot of research on the possible causes of foaming, ranging from wastewater characteristics and composition, operational conditions (e.g., temperature, OLR, HRT, mixing methods, start-up procedures, etc.), biological factors (e.g., bio-surfactants production by microbial lysis, protein denaturation, extracellular polymers substances excreted by microorganisms subjected to certain environmental stress, etc.). Detailed analysis of these can be found elsewhere (e.g., Barber 2005). Where foaming occurs, the causes may be so complex that it may not be cost effective trying to understand the them, many of which may be difficult to change.

The most effective method of foam control is treating the foam directly rather than trying to identify the root causes and implementing changes in the reactor operation and wastewater characteristics. Direct methods for foam control include the following (Speece 2008):

- Chemical methods: Anti-foaming agents (e.g., polydimethylsiloxane), chlorine, etc.
- Mechanical methods: e.g., foam breakers, disintegration methods, water sprays, etc.
- Biological: Enrichment of surfactant utilizing microorganisms
- Ultrasound treatment and pasteurization

The majority of anti-foaming agents have been reported to be very effective and work within 2–10 minutes (Barber 2005).

3

Anaerobic Digestion (AD) of Organic Solid Residues and Biosolids

3.1 Applications, Benefits, and Challenges

Anaerobic treatment of solids and biosolids is commonly referred to as *anaerobic digestion* or simply AD. Depending on the type of waste, AD provides one or more of the following benefits:

a. Solids reduction, through cost effective and sustainable solid waste management and disposal

b. Pollution control, through stabilization of organic solids prior to disposal to avoid uncontrolled decomposition which can result in land, water, and air pollution.

c. Pasteurization of waste for the reduction of pathogens prior to disposal to agricultural land, and to reduce occupational hazard and transmission of disease to animals and humans

d. Bioenergy production

e. Production of organic fertilizer

Process design, operation, and control are relatively simple where (a), (b), and (c) are the principle treatment objectives goals. Where (d) and (e) are also required, detailed process analysis is needed as well as careful consideration of feedstock quality and preparation.

Typical types of *feedstocks* (i.e., solids and biosolids) that can be used in AD are listed in Table 3.1.

Hydrolysis is the limiting step in AD. While some organic solids are easily hydrolysable, others take more time and may require assistance in the form of physical, chemical, and biological pretreatments (covered in Chapter 4). Some of the key considerations for effective AD application, therefore, include the quality and availability of suitable feedstocks, and their pretreatment requirements. Other considerations include management options for the *digestate*, i.e., the semisolid by-product or digested slurry, and availability of skilled manpower.

TABLE 3.1

Typical and Emerging Anaerobic Digestion Feedstocks

Common Feedstocks	Others
• Weeds, wood, grasses, leaves	• Microalgae
• Fruit and vegetables	• Freshwater macrophytes
• Agricultural residues, e.g., residuals from the harvest of cereals (maize, wheat, bailey, sorghum, etc.)	• Seaweed
	• Co-products from other bioenergy technologies, e.g.,
• Co-products of food/beverage processing	○ Pyrolysis oils/syngas
• Organic fraction of municipal solid waste/food wastes	○ Fermentation residues
	○ Extraction residues
• Livestock wastes and slurries	○ Biodiesel residues, etc.
• Biosolids (or sewage sludge, comprising waste activated sludge and primary sludges)	
• Energy crops (i.e., crops cultivated specifically for the generation of bioenergy), e.g., ryegrass, sunflower, maize, sorghum, etc.	

3.2 Mono- and Co-Digestion

Mono digestion is where the feedstock consists of only one type of substrate. Co-digestion involves the digestion of more than one type of feedstock. Co-digestion can improve biogas production when the main source of feedstock is in limited supply, thereby making single feedstock digestion unsustainable. It is also relevant for AD plants located in geographically remote areas where the cost of transportation of the main feedstock could be reduced by supplementing with other types of locally available feedstocks. It can secure the stable year-round operation of anaerobic digesters treating substrates that are seasonal by nature or during crop rotation. Furthermore, co-digestion can contribute to process stability through dilution of constituent toxic compounds contained in any of the co-feedstocks with other co-feedstocks deficient in those compounds. For example, co-digestion can be used in the reduction of microbial ammonia, hydrogen sulfide, and sodium ion toxicity in AD systems. The effectiveness of co-digestion usually lies in the harmonization of parameters between the sources of substrates as illustrated in Figure 3.1.

In considering co-digestion, the following factors must be taken into account:

a. Feedstock availability and quality

b. Cost of obtaining, transferring, preparing (or pre-treating), and storing the extra feedstock(s)

c. Impact on process yields and kinetics of new feedstocks added continuous or intermittently when not available all year round.

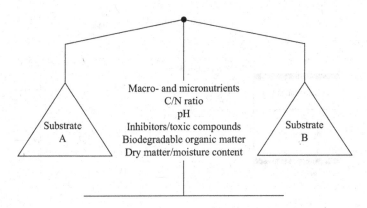

FIGURE 3.1
Co-digestion balance (Adapted from Hartmann et al. 2003).

3.3 Process Variations

3.3.1 Standard Rate Digestion

The standard rate system or *digester* consists of a single-stage operation without mixing. Figure 3.2a is a schematic diagram of a standard rate digester used for biosolids digestion. Stratification may occur within the reactor or digester, however, some mixing occurs in the active zone due to biogas emission and turbulence created during feeding and digestate and supernatant withdrawal. Feeding is normally intermittent. For biosolids AD, the supernatant is normally returned to the inlet of the associated wastewater treatment plant and the digestate sent to a biosolids management facility for further treatment. Standard rate systems are commonly used in biosolids treatment.

3.3.2 High-Rate Digestion

High-rate digesters are designed and operated to maximize biogas. These systems can consist of one or more stages. Figure 3.2b is a schematic diagram of a typical one-stage high-rate biosolids digester. Mechanical mixing enhances contact between substrates and the microorganisms, and thus increases the rate of biodegradation and solids destruction. Up to 50% solids reduction can be achieved in high-rate systems over shorter reaction time and in a smaller reactor volume than in standard rate systems.

In two-stage systems, the second digester is used for digestate storage and settling, and provides further opportunity for the biodegradation of those organic compounds that have not been able to biodegrade sufficiently in the first digester. The second digester may be heated and/or mixed to improve solids destruction. Biodegradation continues in this second digester, at a

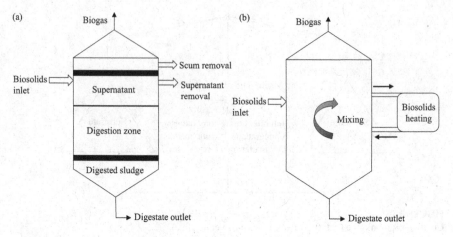

FIGURE 3.2
Schematic diagram of anaerobic digesters (a) standard rate and (b) single-stage high-rate. (Adapted from Metcalf and Eddy 2014.)

much lower rate than in the first digester and this can lead to an overall increase in solids breakdown and biogas yield.

3.3.3 Low-Solids Digestion

AD of feedstocks of between 4% and 10% solids content is referred to as low-solids digestion or *"wet" digestion*. This is the most common AD application. Figure 3.3 shows a typical process configuration of low-solids digestion of organic fraction of municipal solid wastes.

Low-solids feedstock concentration enables easier mixing using mechanical or compressed biogas as illustrated in Figure 3.4. Adequate mixing enhances contact between substrates and microorganisms, and can bring about dilution of toxic compounds where present. Gas production rates by low-solids digestion systems range from 0.5 to 0.75 m^3/kg of biodegradable volatile solids (VS) destroyed (i.e., 0.5–0.75 m^3/kg VS$_{destroyed}$) (Tchobanoglous et al. 1993). One of the disadvantages of the low-solids digestion is the considerable amount of water needed to achieve the desired operational moisture content. This may not be a disadvantage if the feedstocks are of low-solid content, e.g., biosolids. For dryer feedstocks, moisture content and nutrients can be augmented by mixing (i.e., co-digesting) with high moisture content feedstocks such as biosolids, livestock manure, and food and beverage processing wastewaters, where available. Low-solids digestion results in a very dilute digestate, which must be dewatered prior to disposal. Other disadvantages include larger reactor size and space, and greater energy consumption for pumping, mixing, and heating. The handling and disposal of large volumes of the liquid streams resulting from digestate dewatering is also an important consideration in low-solids digestion.

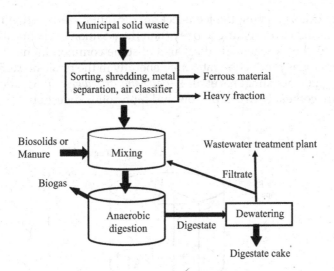

FIGURE 3.3
Typical flow sheet for low-solids digestion of the organic fraction of municipal solid waste.

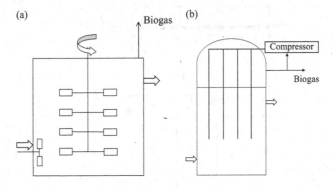

FIGURE 3.4
Low-solids anaerobic digestion systems with (a) mechanical and (b) compressed biogas mixing.

3.3.4 High-Solids (or "Dry") Digestion

High-solids digestion or *"dry" digestion* refers to a digestion process where the solid content in the reactor is greater than 20%. It is characterized by lower water usage, reactor size, space and heating requirements, and higher bio gas production per unit volume of reactor than the low-solids digestion. Mixing is however less effective than in low-solids digestion, and therefore it requires greater attention. Furthermore, the system is less able than low-solids diges-tion to cope with toxicity caused by high ammonia or hydrogen sulfide. Due to poor mixing, adjusting the feedstock mix through co-digestion is a more reliable method of preventing toxicity than the addition of chemicals.

Figure 3.5 shows schematic diagrams of some common high-solid digestion systems. Gas production rates are generally in the range of 0.625–1.0 m³/kg VS$_{destroyed}$ (Tchobanoglous et al. 1993). *'Semi-dry' digestion* occurs in range of moisture content between the low and high solids digestion

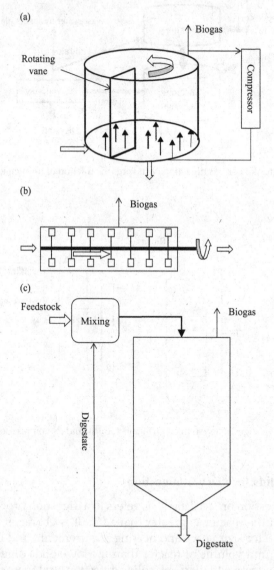

FIGURE 3.5
High-solids anaerobic digestion systems with (a) combined mechanical and compressed biogas mixing, (b) mechanical lateral mixing plug-flow, and (c) plug-flow and recycle.

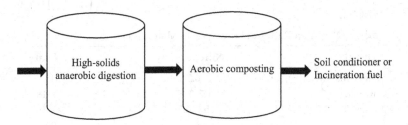

FIGURE 3.6
Flow sheet for combined high-solids anaerobic digestion and aerobic composting.

3.3.5 Combined Anaerobic–Aerobic System

This is a combination of high-solids AD and aerobic composting, in a two-stage system as illustrated in Figure 3.6. The main advantage of this process configuration is the likelihood of a complete stabilization of the biodegradable organic constituents accompanied by a net energy recovery and without the need for major dewatering equipment. Other advantages include enhanced pathogen and volume reduction. The final solid by-product can be used directly as soil conditioner, or as fuel source in incineration plants for electricity and heat production.

3.4 Process Design, Performance, and Operational Control

Process design, performance, and operational control parameters in AD are similar to those presented in Chapter 2 for anaerobic wastewater treatment. However, in AD, the key treatment objective or efficiency indicator is organic solids reduction.

3.4.1 Feedstock C/N Ratio

The ideal C/N ratio for AD ranges from 20 to 30. Table 3.2 shows the C/N ratios of common AD feedstocks. If nutrient supplements are required, nitrogen, for example, can be added by mixing with nitrogen rich substrates such as urea, sewage sludge, livestock manure, etc., and phosphorous by the addition of phosphoric acid or phosphate salt.

3.4.2 Retention Time (RT)

RT refers to the amount of time biodegradable solids fed to the AD system spends in the system. The appropriate reaction time depends on the type of

TABLE 3.2

Carbon/Nitrogen Ratios of Some Waste Materials

Material	% Nitrogen (Dry)	C/N Ratio
Poultry manure	6.3	15
Cow manure	1.7–2.15	14–18
Nightsoil	5.5–6.5	6–10
Urine	15–18	0.8
Blood	10–14	3
Kitchen waste	2.0	25
Seaweed	1.92	19
Water hyacinth	0.9	27
Wheat straw	0.3–0.5	128
Rice straw	0.9	50–60
Rotted sawdust	0.25	200–150
Raw sawdust	0.11	200–500
Food wastes	2–3	15
Total refuse	0.5–1.4	30–80
Wood	0.07	700
Paper	0.2	170
Grass clippings	2–6	12–20
Weeds	2.0	19
Leaves	0.5–1	40–80
Fruit wastes	1.5	35
Biosolids (or Sewage Sludge)		
Activated	5.6	6
Digested	2–4	16

Source: Adapted from Kiely (1997) and Polprasert (2007).

solids and their biodegradability (see Figure 1.4, Chapter 1) and is simplified by the following equation:

$$RT = \frac{V}{Q},$$ (3.1)

where
RT = Retention time, day
V = Reactor volume, m^3
Q = Average feeding rate, m^3/day

It is worth noting that Equation 3.1 is equivalent to HRT (Equation 2.1) for wastewater treatment. However, RT values are generally higher than their wastewater treatment equivalent (i.e. HRT) since (a) RT must take into account the longer time required for hydrolysis of particulate matter, and

TABLE 3.3

Recommended Design Solids Retention Times for High-Rate Anaerobic Digestion of Biosolids

Operating Temperature, °C	Solids Retention Time, day (Minimum)	Solids Retention Time, day (Design)
18	11	28
24	8	20
30	6	14
35	4	10
40	4	10

Source: McCarty (1964, 1968), Metcalf and Eddy (2014).

(b) high-rate wastewater treatment systems separate SRT from HRT to prevent excessive biomass washout, and this is normally achieved by waste sludge recycling or by the use of attached growth systems. These practices allow anaerobic wastewater systems to be operated at low HRT while keeping the SRT relatively high in order to maintain a good amount of microbial biomass within the content.

Too short RT may not allow sufficient time for anaerobic bacteria to metabolize the solid residues, while too long can lead to large reactor volumes, and can result in excessive accumulation of digested materials in the digester. RT generally decreases with increase in digestion temperature, normally starting from 10 days for readily biodegradable solids (e.g., biosolids) in mesophilic digestion, to much longer durations depending on the type of solid residues, types of digester, and operating temperature. Table 3.3 shows the effect of temperature on RT for biosolids digestion. Shorter RT can be applied for higher operating temperature as shown in the table. Optimum RT values for standard rate digesters falls within the range 30–90 days, while much shorter durations are used for high-rate systems.

3.4.3 Solids Loading Rate (SLR)

Solids loading rate represents the amount of organic solids or VS, applied daily per unit volume of the reactor, as expressed in the following equation:

$$SLR = \frac{QS_0}{V},\qquad(3.2)$$

where
  SLR = Solids loading rate, kg VS/m^3·day
  S_0 = Influent volatile solids concentration in feedstock, kg VS/m^3
  V = Reactor volume, m^3
  Q = Average feeding rate, m^3/day

A high SLR will normally result in undesirable excessive volatile fatty acid build-up in the system, which can cause a drop in pH to undesirable levels. On the other hand, very low organic loading may not produce a sufficient amount of biogas to make the operation economically viable. Optimum SLR for standard rate digesters vary from 0.5 to 1.6 VS/m³·day. High-rate systems can cope with higher SLR. In general AD systems can tolerate higher SLR at higher temperatures. Start-up operations usually start with low SLR, which are gradually increased as process stability improves. RT and SLR are important AD design and process control parameters.

3.5 Biogas Production and Operational Criteria

Biogas production rate will depend on several parameters, notably, organic content, reactor type, and operational characteristics (e.g., digestate recirculation rate, temperature, etc.) Table 3.4 shows the effect of reactor type on methane yield in biosolids digestion.

When the composition of the organic content of a feedstock is known, it is possible to evaluate the theoretical methane yield as shown in Table 3.5.

TABLE 3.4

Suggested Design Criteria for Anaerobic Digesters

Parameter	Standard Rate	High Rate
Retention time, day	30–90	10–20
Solids loading rate, kg VS/m³·day	0.5–1.6	1.6–6.4
Digested solids concentrations, %	4–6	4–6
Volatile solids reduction, %	35–50	45–55
Biogas production, m³/kg VS$_{added}$	0.5–0.55	0.6–0.65
Methane content, %	65	65

Source: Peavy et al. (1985).

TABLE 3.5

Theoretical Methane Yield of Typical Organic Compounds

Substrate	Composition	m³ CH$_4$/kg VS
Proteins[a]	$C_5H_7NO_2$	0.496
Lipids	$C_{57}H_{104}O_6$	1.014
Carbohydrates	$(C_6H_{10}O_5)_n$	0.415

Source: Angelidaki and Sanders (2004).
[a] Nitrogen is converted to NH_3.

A generalized methodology for estimating the theoretical methane yield and other useful co-products, such as carbon dioxide, ammonia, and hydrogen sulfide can be obtained from Equations 3.3 and 3.4, with and without sulfate reduction respectively (Rich 1963; Tchobanoglous et al. 1993).

$$C_a H_b O_c N_d + 0.25(4a - b - 2c + 3d)H_2O \rightarrow 0.125(4a + b - 2c - 3d)CH_4$$

$$+ 0.125(4a - b + 2c + 3d)CO_2 \quad (3.3)$$

$$+ dNH_3$$

$$C_a H_b O_c N_d S_e + 0.25(4a - b - 2c + 3d + 2e)H_2O \rightarrow 0.125(4a + b - 2c - 3d - 2e)CH_4$$

$$+ 0.125(4a - b + 2c + 3d + 2e)CO_2$$

$$+ dNH_3 + eH_2S.$$

$$(3.4)$$

For continuously stirred-tank reactor (CSTR) a more simplified version of the above equations has been proposed as shown in the following equations (Gray 2010; Kiely 1997; Tchobanoglous and Schroeder 1987):

$$M_{CH_4} = 0.35(nQC_i - 1.42R_gV) \quad (3.5)$$

Or simply,

$$M_{CH_4} = 0.35(L - 1.42S_t), \quad (3.6)$$

where
M_{CH_4} = Methane production rate, m³/day
n = Fraction of biodegradable COD conversion
Q = Waste flow rate, m³/day
C_i = Influent feedstock COD concentration, kg/m³
R_g = Microbial growth (or volatile solids accumulation) rate, kg/m³·day
1.42 = Conversion factor from microorganisms to biodegradable COD
V = Reactor volume, m³
$L = nQC_i$ = Biodegradable COD removed, kg/day
$S_t = R_gV$ = Volatile solids (VS) produced, kg/day
S_t can be estimated from the following equation:

$$S_t = aL/(1 + bt_s), \quad (3.7)$$

where
a = mass of volatile solids produced per kg of biodegradable COD removed
b = endogenous respiration constant
t_s = retention time (RT), day

TABLE 3.6

Biogas Potential of Some Feedstocks

Feedstock	Biogas yield (m³/tonne-VS)
Manure	
Cow	200–400
Pig	350–550
Chicken	350–550
Agricultural Residues	
Maize (whole crop)	660–1000
Grass	500–750
Sorghum	490–620
Potatoes	460–650
Sugar beet	380–600
Industrial Wastes	
Pulp and paper making	400–800
Fermentation, brewery, and distilling	400–800
Food processing	400–600
Abattoir wastes	550–1000
Fat and oils	600–1300
Municipal (segregated) wastes	400–500
Market (segregated) wastes	500–600
Biosolids (or sewage sludge)	250–350

Source: Adapted from German Solar Energy Society (DGS) and Ecofys (2005) and IEA Bioenergy (2011).

An approximation of Equation 3.5 ignoring growth rate is given in Equation 3.8, and can be used for initial estimation of methane production rate.

$$M_{CH_4} \approx 0.35 Q C_i. \tag{3.8}$$

The rate of biogas yield per unit weight of organic wastes can vary widely depending on the characteristics of feedstock and operating conditions. Table 3.6 shows range of biogas yield for various types of feedstocks.

3.6 Modes of Operation

3.6.1 Batch Operation

Batch systems are single-stage systems operated intermittently. The digester is filled once with fresh substrate and allowed to go through all the biodegradation steps until such a time that it is deemed suitable to remove the digestate and replaced with fresh feedstock. Biogas production rate increases over time until it reaches a maximum and begins to decrease. The rate of biogas production is used as an indicator of when to remove the digestate. Up to 50% of the digestate in the reactor can be removed and replaced for each round of treatment, the remaining serves as seed for the next batch. Batch systems are more commonly used in the laboratory scale determination of the biochemical methane potential of different substrates. They are low cost systems, both in terms of capital and operational costs, but require large space, and hence commercially suitable in rural and remote areas where land is more readily available. They are also widely used in small-scale solid waste and slurry digestion in developing countries.

3.6.2 Semi-Continuous Operation

Where there is a steadier supply of feedstock, batch operation can be replaced with semi-continuous operation, which involves more frequent feeding and withdrawal of digestate. The amount of feedstock added and removed depends on the nature of waste, and generally smaller than in batch operation. However, more attention is required in system operation.

3.6.3 Continuous Operation

Feeding and withdrawal of digested slurry take place simultaneously. These operations involve pumping and mixing which require external energy supply; hence, it is more suitable for low-solids digestion. The risks of process failure are greater, therefore closer attention is required in process monitoring and control.

4

Pretreatment in Anaerobic Treatment

4.1 Need for Pretreatment

Organic constituents of anaerobic feedstock may include monomers, natural polymers (such as starch, lipids, elastin, collagen, keratin, chitin, and lignocelluloses), and synthetic polymers (such as polyesters, polyethylene, and polypropylene). Monomers are relatively easy to breakdown under normal anaerobic process conditions; however, polymers take a longer time to break down as shown in Figure 1.4 due to their chemical structures therefore may require pretreatment to improve their amenability to biodegradation. Lignocellulose polymers are usually the major organic constituents of municipal and industrial residues.

Lignocelluloses consist mainly of three polymers, viz., cellulose, hemicellulose, and lignin. Cellulose in a plant is composed of both parts with organized or crystalline structure and parts with less organized or amorphous structure. The cellulose chains are weakly held together by hydrogen bonds to form cellulose fibrils or cellulose bundles (Hendriks and Zeeman 2009; Laureano-Perez et al. 2005; Taherzadeh and Karimi 2008). This structure makes cellulose resistant to both biological and chemical treatment.

Hemicellulose is a complex carbohydrate structure that consists of different polymers such as pentoses, hexoses, and sugar acids and has lower molecular weight than cellulose. It serves as a connection between cellulose and lignin fibers and provides more structural rigidity in the cellulose–hemicellulose–lignin network (Hendriks and Zeeman 2009; Laureano-Perez et al. 2005). The solubility of some hemicellulose compounds increases with increase in temperature and decreases in the following order: mannose, xylose, glucose, arabinose, and galactose. The dominant hemicellulose is dependent on the source of the organic matter. For example, xylose is dominant in hardwoods and agricultural residues such as grasses and straw, while mannose is dominant in softwoods (Hendriks and Zeeman 2009; Saha 2003; Taherzadeh and Karimi 2008). Hemicelluloses are generally less resistant than cellulose to hydrolysis to their monomer constituents.

Lignin is one of the most abundant polymers in nature, and a component of the plant cell wall. It has an amorphous heteropolymer, and consists of three different phenylpropane units held together by different types of linkages to form three-dimensional structures, with the nature of the latter determining its solubility in acid, neutral, or alkaline environments. The amorphous heteropolymer is generally insoluble in water and optically inactive (Hendriks and Zeeman 2009). All these make lignin difficult to biodegrade. It is the most recalcitrant component of the plant cell wall, and the greater its proportion in any feedstock, the greater the feedstock resistance to chemical and biological degradation.

4.2 Mechanical Pretreatment

4.2.1 Collection and Segregation

Waste collection method and frequency, and waste segregation affect the quantity and quality of municipal solid wastes (MSW). MSW segregation at source or at waste reception centers increases the organic fraction of municipal solid wastes (OFMSW). The use of OFMSW in anaerobic digestion, as opposed to the entire MSW, ensures that the digester is not fed with inorganic and non-readily biodegradable materials such as plastics, wood, bones, leather and textile materials, broken bottles, etc. The effectiveness of waste collection and separation is dependent on the level of public awareness, available infrastructure, and technical capacity, and these in turn are dependent upon the level of affluence of the waste producers. In some affluent communities, where organic waste conversion and reuse is part of the waste management strategy, separate collection facilities for food and garden wastes are used available for individual homes, institutions, hotels, etc. This method of OFMSW collection is referred to *source-separation*. Source-separation is often complemented by the use of manual or mechanical separation systems at waste reception centers.

4.2.2 Size Reduction

Size reduction increases the surface area of the feedstock in contact with microorganisms during the digestion process. Also, the smaller the particle size, the greater the solubility of the substrates and hence, the greater the rate of hydrolysis. It has been reported that the rate of biodegradation and thus biogas production is inversely proportional to the average diameter of the substrate (Hills and Nikano 1984; Sharma et al. 1988). Mechanical particle size reduction incurs additional capital and operational expense, the latter in the form of energy, equipment maintenance and labor. In some cases, heat co-treatment can increase the effectiveness of mechanical size reduction operations. Although the smaller the particle size the greater the overall cost; a particle size of 1–2 mm is considered as the most cost effective for hydrolysis of lignocellulose substrates (Schell and Hardwood 1994). The effectiveness of size reduction in bringing about a net increase in biogas yield depends on the type and characteristics of the feedstock (Kuzmanova et al. 2018). A detailed economic analysis is, therefore, required prior to employing any size reduction techniques in order to ascertain and quantify the potential economic benefits.

4.2.3 Ultrasound (US)

Sound frequency above 18 kHz is considered to be ultrasound (US). When US is applied to an aqueous solution or suspension an increase in mixing, shearing, and mass transfer takes place. Under certain conditions, cavitation is produced, which manifests itself in the production of tiny bubbles that implode to produce so-called hotspots which tend to generate highly reactive hydroxyl radicals and physical and chemical transformations (Bremner 1990). Cell disintegration methods based on US and temperature-pressure induced pulping have been successfully used to increase the biogas yield in the digestion of biosolids (Neis et al. 2001; Nickel and Neis 2007; Onyeche et al. 2002; Pérez-Elvira et al. 2010; Pilli et al. 2011; Show et al. 2007; Tiehm et al. 1997, 2001). US treatment tends to destroy the cell floc and granular structure, and fragment the cell walls into low particle sizes in addition to releasing soluble cell materials that are readily biodegradable. Cell walls often create physical barriers to enzymes of hydrolytic microorganisms; thus, removing them can bring about a general increase in the rate of biodegradation and boost biogas production. US is not very effective in the pretreatment of lignocellulosic materials (Onyeche et al. 2002); however, it has been reported to increase the hydrolysis of cellulose (Zhang et al. 2013). Detailed information on the application of US in biological processes can be found elsewhere (e.g., Kwiatkowska et al. 2011).

4.3 Biological Pretreatment

4.3.1 Anaerobic Processes

This is essentially a two-stage system where hydrolysis and acidogenesis takes place in the first stage and methanogenesis in the second stage. A two-stage system allows hydrolysis and acidogenesis to be operated at optimum pH and temperature ranges of 4–6 and 30°C–50°C, respectively, which are suitable for the hydrolysis of a wide range of readily and non-readily biodegradable organic substrates.

Separation of hydrolysis/acidogenesis from the methanogenesis also enables separate recovery of methane-rich biogas produced in the methanogenesis stage from carbon dioxide (sometimes and hydrogen sulfide-rich) biogas produced in the hydrolysis/acidogensis stages. Shanmugam and Akunna (2008) and Nizami et al. (2012) recovered biogas with 86% and 71% methane, respectively, from the methanogenic zones of multi-stage anaerobic treatment systems. Anaerobic pretreatment also lowers the effect of inhibitory compounds (e.g., sulfides, ammonia, phenols, etc.) on the more sensitive methanogens. In practice, anaerobic pretreatment can be combined with other types of pretreatment.

4.3.2 Aerobic Composting or Digestion

Aerobic microorganisms can generally produce lignocellulose-degrading enzymes more rapidly and in larger amounts than anaerobic microorganisms. This aerobic property can be utilized to quicken the hydrolysis of solids to their soluble constituents such as carbohydrates, amino acids, and volatile acids, which can subsequently be converted to biogas in an anaerobic digester. Moreover, some aerobic organisms can break down compounds that might otherwise inhibit methanogenesis.

Aerobic composting can be carried out in mesophilic or thermophilic temperatures, and involves the actions of a wide range of mesophilic and thermophilic bacteria. Thermophilic aerobic digestion pretreatment, carried out at a temperature range of 65°C–85°C (with 80°C as optimum), is an essential part of anaerobic digestion of biosolids in some countries to achieve pasteurization where the digestate is used as fertilizer for food crops. Thermophilic aerobic digestion of biosolids followed by mesophilic anaerobic can lead to up to 1.5 times increase in biogas yield with higher methane content, and produce digestate with improved dewaterability. However, it can lead to a decrease in biogas yield where the aerobic digestion is carried out over a long period due to excessive aerobic organic conversion to carbon oxidation. Aerobic pretreatment should ideally be limited to a very short duration; otherwise, significant amount of organic substrates that would have been converted to biogas may be lost to carbon dioxide. Prior laboratory scale trials are therefore required before full-scale application in order to establish operational conditions that could prevent avoidable organic carbon losses in the aerobic stage. Another important by-product of aerobic composting is heat. With proper design, this heat can be harnessed for heating the anaerobic digester.

4.3.3 Fungi

Fungi pretreatment can lead to biodegradation of recalcitrant organic compounds. For example, fungi can break down lignin contained in straw (Geib et al. 2008; Ghosh and Bhayyacharyya 1999; Mustafa et al. 2017). White-rot fungi have been known to breakdown compounds that are toxic to methanogens, and those toxins that can affect agricultural reuse of digestate. Other successful applications include fungi detoxification of phenolic compounds

(Dhouib et al. 2006; Hodgson et al. 1998) and of coffee cherry husks (Jayachandra et al. 2011) prior to anaerobic digestion. Most of the reported applications of fungal pretreatment in anaerobic digestion are still in experimental stages.

4.3.4 Enzymatic Hydrolysis

Enzymatic hydrolysis occurs naturally in anaerobic digestion systems. Microorganisms can synthesize and secrete different enzymes that hydrolyze proteins, carbohydrates, and lipids to simpler products, which are subsequently converted to biogas. Some of these enzymes are commercially available and can be used as artificial supplements in anaerobic systems to enhance hydrolysis and biodegradation in general.

Enzyme pretreatment can be carried out in three ways depending on the system configuration:

a. Direct addition of enzymes into a single reactor system

b. Addition of enzymes into the hydrolysis/acidogenesis reactor of a multi-stage system

c. Addition of enzymes to a dedicated enzyme pretreatment vessel

Operation (a) will require a relatively high quantities of enzymes than the others, and (c), although most effective of all, requires higher capital and operational costs due the extra reactor vessel. Operation (b) is ideal in multi-stage anaerobic treatment systems.

Enzymatic pretreatment has been successfully used in anaerobic treatment of wastes from dairies, brewery, distilleries, and abattoirs (Al-Alm Rashed et al. 2010; Cammarota et al. 2001; Jung et al. 2002; Mallick et al. 2010a, b; Massé et al. 2001, 2003). It can bring about up to 40% increase in volatile solid reduction in biosolids digestion. Mallick et al. (2010a, b) observed up to 87% increase in soluble COD production in enzymatic pretreatment of distillery spent wash centrate and pot ale containing intact yeast cells.

The effectiveness of enzymatic hydrolysis depends on the nature and composition of substrates, and the types, combinations, and dosages of various contributing enzymes. Some of the available commercial enzymes are very expensive; therefore, detailed economic analysis, as well as comparative cost analysis with other types of pretreatment should be carried out prior to process selection. Enzymatic pretreatment can be combined with other pretreatment methods presented in this chapter.

4.3.5 Bio-Augmentation

Bio-augmentation is a process where supplementary microbial cultures are added to a biological system to facilitate biodegradation. Strictly speaking, bio-augmentation is a co-pretreatment action, which enhances the activities of microorganisms already present in the system. Some researchers have reported improvements in biogas yield by bio-augmentation through increase in the rates of hydrolysis of cellulose, hemicellulose, and lignin, and through increase in the diversity of the methanogens. These observations have been based on the use of pure microbial strains under controlled laboratory conditions (Nielsen et al. 2007; Ozbayram et al. 2016; Peng et al. 2014; Tsapekos et al. 2017; Zhang et al. 2015a). Data on full-scale applications is still scanty. There are many commercial products claiming to significantly increase the biogas yield through bio-augmentation. While these claims may be correct, it is important to note that anaerobic treatment processes rely on the activities of different groups of mutually dependent microorganisms. The increase in activities of only some of these groups may not necessarily produce the desired results. Moreover, operational factors such as organic loading rate, temperature, pH, and microbial inhibitors (e.g., O_2, H_2S, NH_3) can play a vital role in determining the balance and fate of the participating microbial groups within the system.

4.3.6 Bio-Supplementation

Bio-supplementation or bio-stimulation refers to addition of trace metals or micronutrients to stimulate microbial metabolic processes. Metals such as Fe, Co, Mo, Mn, Se, Cu, and Ni when available in relatively small amounts can stimulate methanogenic activities. Certain feedstocks are naturally deficient in some of these metals and their addition may be necessary to ensure process stability. The use of trace metal supplements to boost biogas production has proved effective, particularly in the digestion of food wastes (Banks et al. 2012; Demirel and Scherer 2011; Facchin et al. 2013; Zhang et al. 2015b).

4.4 Chemical Pretreatment

4.4.1 Acid and Alkaline

Mild acid and/or alkaline pretreatment can improve solids solubilization and can be effective in the reduction of pathogens. The key design and operational considerations include solids concentration and nature (e.g., biosolids, lignocellulose, etc.); treatment time and temperature; alkali or acid type and concentration; and preceding and follow-up treatment. With appropriate design, chemical pretreatment is highly effective and can lead to more than 50% increase in solids reduction and biogas yield. However, most of the reported studies have been based on laboratory and pilot scale applications.

Acid-pretreatment disrupts and breakdowns structural bonds in lignocellulosic substrates (Knappert et al. 1981). It is normally carried out at pH ranges of 4–5 using nitric acid (HNO_3), at room or high temperatures. Other acids such as hydrochloric acid (HCl) and sulfuric acid (H_2SO_4) can also be used; however, chemical reactions with these acids can result in the formation of recalcitrant compounds and microbial inhibitory by-products, such as furfural and hydroxymethylfurfural (HMF) (Ariunbaatar et al. 2014a). Moreover, addition of H_2SO_4 can introduce excessive amount of sulfate to the system that can bring about hydrogen sulfide toxicity. In general, strong acidic pretreatment should be avoided as they are more likely to form undesirable by-products, and can also bring about low biogas yield due to excessive oxidation and loss of biodegradable COD. High costs of chemicals and post-treatment neutralization necessary for the follow-up anaerobic digestion should be taken into account in assessing the suitability of acid pretreatment.

Alkali pretreatment is more commonly used than acid pretreatment due to its compatibility with anaerobic digestion process since it increases the alkalinity of the substrate and consequently able to cancel out low pH resulting from the acidogenesis. It acts by causing swelling of lignocellulose and partial solubilization of lignin (IEA Bioenergy 2014; Kong et al. 1992). Sodium hydroxide (NaOH) is commonly used. Others are potassium hydroxide (K(OH)), magnesium hydroxide ($Mg(OH)_2$), and calcium hydroxide or lime ($Ca(OH)_2$). Like acid, alkali treatment can lead to the formation of undesirable by-products. For example, excessive use of metallic hydroxides can lead to metal toxicity (see Table 1.4). Moreover, some recalcitrant compounds can be formed, such as aromatic substances, from the treatment of lignocellulosic substrates.

Chemical pretreatment can be combined with thermal and enzymatic treatment. Further literature can be found elsewhere (e.g., Ariunbaatar et al. 2014a; Delgenès et al. 2003; IEA Bioenergy 2014; Montgomery and Bochmann 2014; Obata et al. 2015; Speece 2008).

4.4.2 Ozonation

Ozone (O_3) is a powerful oxidizing agent, more powerful than oxygen and as effective as hydrogen peroxide in oxidizing a wider range of organic compounds. Ozonation involves two modes of action: (a) direct reaction of the ozone on the solids and (b) indirect action on the solids by hydroxyl radicals formed when ozone is dissolved in water (Brière et al. 1994). The latter causes a rise in pH. At high pH, the radicals are more dominant and directly oxidize soluble organics. On the other hand, at low pH, ozone is more dominant and directly attacks the particulate fractions.

Oxidative pretreatment degrades lignocellulose compounds in the same manner as alkaline pretreatment, and it can also break down lignin. Unlike acid and alkali treatment, ozone treatment does not bring about increase in salt concentration, and can destroy or convert recalcitrant compounds to more biodegradable forms. It can also reduce the presence of enteric pathogens in the feedstock and is effective in the reduction of odor and color. Ozone has the advantage of being relatively easy to be produced on site, thus less susceptible to expensive transportation and storage requirements. Ozonation can, however, lead to the formation of microbial inhibitory substances caused by excessive dosage or by its reactions with certain compounds, such as organic peroxides, H_2O_2, low molecular-weight alcohol, some carboxylic acids, and aldehyde (Gilbert 1983). This drawback can be overcome by appropriate digester start-up procedures that encourage gradual microbial acclimatization of the pretreated substrate (Narkis and Schneider-Rotel 1980). Most of the applications reported in the literature have been carried out with biosolids and industrial wastewaters; although, successful applications with food wastes have also been reported (Ariunbaatar et al. 2014b). Doses of up to 3.4 g O_3/L of biosolids and 30 mg O_3/g VS have been reported to bring about increases in solids reduction, biogas yield, and biodegradability of up to 50, 50%, and 60%, respectively.

4.5 Thermal

High temperatures in the presence of water can bring about disruption of structural bonds present in cellulosic and lignocellulosic substrates. Thermal treatments can be combined with chemical and/or mechanical treatment. The three types of thermal pretreatment processes commonly used in anaerobic digestion are as follows:

a. High temperature treatment, in the range of 100°C–200°C without pressurization,

b. High temperature and pressure treatment or *wet air oxidation* usually in temperature and pressure ranges of 200°C–377°C and 5–15 MPa (or 49.4–148 atm), respectively, and

c. Pyrolysis, at temperatures above 400°C

Thermal pretreatment also brings about waste pasteurization, which is necessary where the digestate is used as fertilizer for food crops.

4.5.1 High Temperature

This is commonly employed in digestion of biosolids, where high temperatures cause rupturing of microbial cells releasing soluble COD. The level of solids reduction and increase in biogas yield depends on feedstock composition (e.g., relative proportions of primary and waste activated biosolids), solid content, and treatment temperature and residence times. Solid solubilization increases with increase in temperature and reaction time, with typical values of these parameters being in the range of 170°C–180°C and 1–2 h, respectively. Beyond this temperature range, the process may lead to a decrease in soluble organic solids due to polymerization, i.e., formation of larger molecules through the reaction of low sugars and amino acids, and formation of poorly biodegradable substrates. Some authors have observed a significant improvement in methane production in anaerobic digestion of food waste pretreated at a lower temperature and reaction time of 80°C and 1.5 h, respectively (Ariunbaatar et al. 2014b). With proper process design, increases of up to 40% and 60%, respectively, of solids reduction and methane yield can be obtained in the treatment of waste activated biosolids. However, the treatment is less effective for non-microbial biomass such as OFMSW and primary biosolids from municipal wastewater treatment. Another disadvantage is its relatively high-energy requirement.

4.5.2 Wet Air Oxidation

Wet air oxidation takes place in pressurized vessels, at pressures and temperatures of typically above 13.8 MPa (136 atm) and 288°C, respectively, for periods of typically less than 30 min. It can lead to partial or total oxidation of all types of organic compounds present in the feedstock. Treatment efficiency increases with increase in temperature and reaction time. However, care must be taken to ensure that total organic carbon oxidation does not take place which will bring about lower biogas yield. When used for pretreatment of biosolids, it can lead to up to 1,000% increase in solubilized COD, and producing biogas of up to 90% CH_4 methane content (Song et al. 1992). With energy crops, wet oxidation at lower temperature and pressure ranges of 170°C–200°C and 20–30 atm respectively, has been reported to result in 20%–30% increase in biogas yields at significantly shorter residence times (IEA Bioenergy 2014). Wet air oxidation can also be used to destroy toxic and refractory organic compounds present in some industrial wastes (e.g., forestry, pharmaceutical, landfill leachate, etc.), such as phenolic compounds, aromatic hydrocarbons, carboxylic acids, etc. or convert them to their more readily biodegradable monomers.

Where the waste is made up of large amounts of hazardous and toxic compounds, partial oxidation followed by anaerobic digestion may not be effective in detoxifying the waste. A complete thermal oxidation by the *super-critical water oxidation* process may be considered. Super-critical water oxidation takes place at temperatures and pressures above 374.15°C and 22.1 MPa (218 atm), respectively. High destruction efficiencies (>99.9999%) over relatively short residence times (in the order of seconds to minutes) are achievable due to relatively high solubility of organic compounds and oxygen under these conditions.

Wet or super-critical oxidation processes are very capital and energy intensive; hence, a detailed cost benefit analysis must be carried out prior to industrial scale application.

4.5.3 Pyrolysis

Pyrolysis involves heating the feedstock to over 400°C in a closed vessel without oxygen. This will cause the breakdown of organic solids to the following fractions:

a. Gas stream (or syngas) composing primarily of hydrogen, methane, carbon monoxide, carbon dioxide, and others, depending on the characteristics of the original feedstock.

b. Liquid, consisting of tar and/or oil, containing high value chemicals such as acetic acid, acetone, methanol, and complex oxygenated hydrocarbons.

c. Solid, consisting of char which is an almost pure carbon and any inert material contained in the original feedstock.

For example, if the feed material is pure cellulose, Equation 4.1 represents the pyrolysis reaction (Tchobanoglous et al. 1993):

$$3\underset{\text{cellulose}}{(C_6H_{10}O_5)} + \text{heat energy} \rightarrow 8H_2O + \underset{\text{tar/oil}}{C_6H_8O} + 2CO + 2CO_2 + CH_4 + H_2 + \underset{\text{char}}{7C}.$$

$$(4.1)$$

All these by-products are different forms of fuel. Pyrolysis oils can be used directly as fuel or upgraded to fuels with higher calorific values or high value chemicals. Since lignocellulose polymers are the major organic constituents of woody materials and municipal wastes, oils produced from their pyrolysis are normally high in water content, which reduces their calorific value. On the other hand, oils derived from pyrolysis of scrap vehicle tires and plastics are composed mainly of carbon and hydrogen, and therefore, have higher calorific values. Similarly, the composition of the syngas depends on the type and nature of the original feedstock. Syngas produced from pyrolysis of mainly lignocellulosic feedstocks contains relatively high amounts of carbon dioxide and carbon monoxide, and therefore, generally have low calorific value. While syngas from feedstock with less oxygenated constituents such as scrap tires contains high concentrations of hydrogen, methane, and other hydrocarbons, and have high calorific value.

The low quality oils and syngas, which are not suitable for direct use as fuel for mechanical engines, can be upgraded to biogas through anaerobic digestion. Table 4.1 shows soluble COD concentrations of oils obtained from pyrolysis of various types of solid residues at different temperatures

TABLE 4.1

Effect of Pyrolysis Temperature and Time on Oil COD Concentrations

Feedstock	Temperature (°C)	Heating Time (min)	COD of Oil (g/L)
Softwood pellet	700	10	760
Softwood pellet	550	10	792
Softwood pellet	550	9	660
Wheat straw pellet	550	5	526
Wheat straw pellet	700	6	459
Food waste digestate	550	10	513
Seaweed species	350	20	400

Source: Akunna et al. (2015).

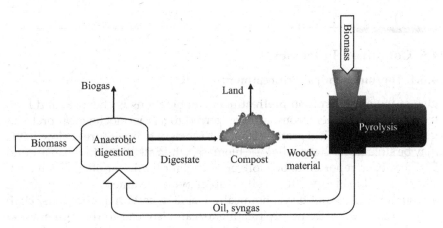

FIGURE 4.1
Combined anaerobic digestion and pyrolysis of municipal wastes or woody biomass.

and heating (or residence) times. The oils can be used as co-substrate in co-digestion systems. The use of these oils as substrate in anaerobic systems is still in early stages of research and development; however, initial results have identified microbial acclimatization as the main process limiting requirement for effective operation (Akunna et al. 2015; Guiot et al. 2011, 2013; Hübner and Mumme 2015).

Figure 4.1 shows a conceptual schematic diagram of combined anaerobic digestion and pyrolysis processes for municipal and woody biomass.

4.5.4 Microwave (MW) Irradiation

Microwave (MW) pretreatment uses heat generated from extensive intermolecular collision to effect and accelerate physical, chemical, and biological changes in the feedstock. It can be applied sole or combined with other types of pretreatment discussed above. For example, MW pretreatment of alkaline-treated glycerol has been reported to improve hydrolysis (Diaz et al. 2015). The performance efficiency of MW pretreatment, however, depends on the nature of the organic constituents. For example, Diaz et al. (2015) reported 12% and 30% reductions in lignin content in MW pretreatment of rice husk and corn straw, respectively. The pretreatment has also been reported to achieve a mere 20% increase in biogas yield for waste activated biosolids (Houtmeyers et al. 2014), which might not be enough to compensate for the energy utilized for the process. MW pretreatment process is a low energy technology and is generally simple to operate when compared with high temperature processes. Its application is, however, still in early stages of research and development.

4.6 Combined Processes

4.6.1 Thermochemical Pretreatment

One of the most common pretreatment combinations is chemical and high temperature treatment, community referred to as thermochemical pretreatment. Thermochemical pretreatment with acids at temperatures below 160°C may be suitable for lignocellulosic biomass, while alkali treatment at around 50°C has been reported to be more beneficial for lignin- or phenol-rich feedstocks (IEA Bioenergy 2014). Follow-up enzyme treatment is also possible, although this may not be economically viable in certain applications. High temperature and extreme pH treatments can also lead to the formation of significant amounts of refractory COD. Prior laboratory and pilot scale optimization studies are, therefore, required before commercial scale operations.

4.6.2 Thermomechanical Pretreatment

Thermomechanical pretreatment combines mechanical and high temperature treatments. Examples include *grinding and heating* and *steam explosion*. Grinding and heating is often associated with moderate gains in terms of solids reduction, while steam explosion has proved to be economically viable with certain feedstocks.

Steam explosion is a combination of two processes: (a) autohydrolysis of the polymeric components caused by the organic acids generated in situ and (b) physical disintegration of the lignocellulosic matrix when it explodes on decomposition (Delgenès et al. 2003). The process is carried out by the addition of high temperature and pressurized steam into a closed vessel (or reactor) containing the feedstock to achieve autohydrolysis, and subsequently releasing the mixture through a small orifice to cause a sudden pressure drop to bring about a strong shear effect that further aids the disintegration of particulate matter. Treatment temperatures and pressures are usually in the ranges of 180°C–260°C and 1–4 MPa (10.2–40.8 atm), respectively, and the process duration varies from a few to several minutes (Sun and Cheng 2002). When applied for pretreatment of biosolids, some proprietary steam explosion products claim to be capable of achieving more than 100% increase in biogas yield, up to 70% solids destruction and increase in the dewaterability of the digestate by up to 50%. Lower efficiencies have been reported with feedstocks that do not contain significant quantities of microbial cellular materials such as straw (Risberg et al. 2013; Vivekanand et al. 2012).

4.6.3 Extrusion

This involves forcing the feedstock through a tube which is subjected to high pressures, temperatures, and shear forces and subsequently ejecting the pressurized and hot feedstock from a small orifice in order to bring about a sudden pressure drop to disintegrate the substrate structure. Some authors have observed improved biogas yield due to this pretreatment under laboratory conditions (Hjorth et al. 2011), however, commercial scale applications may be hampered by high equipment and energy costs.

4.7 Summary of Common Pretreatments

Table 4.2 shows a comparison of some of the pretreatments discussed in this chapter. More information on the effect of different pretreatments on the breakdown of lignocellulose can be found elsewhere (e.g., Hendriks and Zeeman 2009; Taherzadeh and Karimi 2008).

TABLE 4.2

Advantages and Disadvantages of Some Common Pretreatment Technologies

Pretreatment Methods	Advantages	Disadvantages
Mechanical (sorting, separation, shredding, and milling)	• Improves feedstock quality and digestate value • Increases surface area for microbial action • Can be cheap and easy to implement • Low energy consumption • May not require highly skilled manpower • Essential initial pretreatment before application of other pretreatments	High energy requirement, equipment, and maintenance costs, particularly where high feedstock quality and very small particle sizes are required

(Continued)

TABLE 4.2 (*Continued*)

Advantages and Disadvantages of Some Common Pretreatment Technologies

Pretreatment Methods	Advantages	Disadvantages
Ultrasound	Effective for cellular biomass such as secondary biosolids (sludge) hydrolysis	• Effective for a narrow range of applications • High capital and operational costs • Requirement of highly skilled manpower
Anaerobic	Easy and cheap to setup and operate	• Slow and requires large footprint • Little impact on non-readily biodegradable and recalcitrant organics
Aerobic	Easy and cheap to setup and operate	• Requires large footprint • Risk of losing significant amount of organic carbon to carbon dioxide • Little impact on non-readily biodegradable and recalcitrant organic • Odor and flies nuisance if carried out in open spaces
Enzymatic	• Effective for less heterogonous wastes, e.g., biosolids (or sewage sludge) • Low energy requirement	• High cost of enzymes • Difficulties in selecting appropriate enzymes and dosages • Limited full scale application and experience
Chemical	• Alkali addition can provide buffer • Alkali and concentrated acid degrade lignin • Dilute acid degrades hemicellulose	• Cost of chemicals, and their storage and handling costs and risks • Risk of production of recalcitrant compounds • Risk of toxicity from metal ions (e.g., from sodium hydroxide) and sulfate (if sulfuric acid is used) • Acid can cause corrosion problems
Thermal treatment	• Generally more effective for cellular biomass such as secondary biosolids (or sludge) hydrolysis • Steam explosion degrades both lignin and hemicellulose	• High energy, equipment, and maintenance costs • Higher costs for applications on non-cellular biomass

Source: Adapted from Taherzadeh and Karimi (2008), Hendriks and Zeeman (2009), IEA Energy (2014).

4.8 Assessing the Effects of Pretreatment

Most pretreatments involve additional capital and operational costs, which vary widely depending on the method(s) employed. Their effectiveness depends on feedstock characteristics and the intended management options for the digestate. A comprehensive cost-benefit analysis is therefore required when considering potential applications. Some of the criteria used in assessing their performances are explained below.

4.8.1 Chemical Analysis

Soluble or settled chemical oxygen demand (SCOD) or total organic carbon (TOC), total or volatile solids, and ammonia or total Kjeldahl nitrogen (TKN) (i.e., sum of nitrogen in ammonia and ammonium, and nitrogen bound in organic substrates) content of the pretreated feedstock are the key parameters used to assess the effectiveness of pretreatment operations and/or processes in anaerobic process applications. The solids content also informs on the quantity and nature of the solids that will eventually form part of the digestate.

The SCOD test measures the total organic content, which is oxidizable by a strong oxidizing agent, while the TOC measures the total concentration of carbon as organic compounds. The SCOD and TOC measurements do not inform on the biodegradability of the pretreated feedstock. Although their values can be used to obtain the theoretical methane yield, the actual methane yield will depend on the biodegradability of the pretreated substrate. They do not also indicate whether or not the substrate contains microbial inhibitory compounds.

4.8.2 Biochemical Methane Potential

The biochemical methane potential or biomethane potential (BMP) tests assess the affinity of anaerobic microorganisms to substrates intended for anaerobic treatment. In general, BMP tests inform on the following:

a. Level of substrate biodegradability
b. Rate of biodegradation
c. Methane yield
d. Digestate characteristics

The tests are carried out in anaerobic microbial culture vessels of capacity varying typically from 100 to 2,000 mL, in optimum nutrient, microbiological, and mesophilic conditions over a period of usually not less than 30 days.

Detailed description of BMP tests and analysis can be found elsewhere, e.g., Owen et al. (1979), Speece (2008), Hansen et al. (2004), Angelidaki et al. (2009), Rosato (2018), etc.

Figure 4.2 shows a typical set of BMP test results and their possible implications. The figure indicates that the outcomes are dependent on test duration and the source of inoculum, with the latter capable of the rate of biodegradability. Inoculum from sources treating similar substrates can bring about early and rapid metabolic activities, while those from other sources may require longer acclimatization period. Extended test durations tend to reduce the effects of the latter.

Figure 4.2 identifies three typical outcomes of BMP tests. They are as follows:

i. Similar results as those from none pretreated feedstock, as illustrated in Figure 4.2a. Here, pretreatment will not bring about any additional benefits.

ii. Increase in the rate of biodegradation, as illustrated in Figure 4.2b. Here, pretreatment can lead to significant economic gains, through either a reduction in reactor size and consequent decrease in capital and operational costs, or an increase in the amount of feedstock treated for the same size of reactor (as in none pretreated feedstock), and hence, increased biogas production.

iii. Increase in biogas yield over a longer duration, as illustrated in Figure 4.2c, due to any or all of the following:

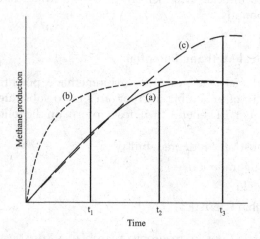

FIGURE 4.2
Typical BMP results showing (a) maximum methane yield without pretreatment at retention time close to t_2, (b) pretreatment results in increase in biodegradation rate and shorter retention time, t_1, and (c) pretreatment leads to increase in methane yield over a longer retention time, t_3 (Adapted from IEA Bioenergy 2014).

- pretreatment leads to the conversion of refractory organic constituents to forms that are amenable to microbial biodegradation, and/or,
- pretreatment causes the formation of organic compounds that require longer microbial acclimatization periods for their biodegradation.

Overall, pretreatment can increase biogas yield. However, this may require increase in reactor sizes and associated operational requirements e.g., increase in heating. It is also important to note that levels of microbial acclimatization that can occur in BMP tests are generally much lower than levels obtainable in full-scale plants that operate continuously over much longer periods.

The BMP assay does not always accurately predict full-scale performance due to the following reasons:

a. They are conducted under optimal conditions of nutrients, temperature, and mixing. Many of these operating parameters cannot be achieved with the same level of accuracy in pilot or commercial scale systems.

b. Being batch tests, there are no regular input of fresh substrates into the BMP test bottles during the test duration meaning that hydrolysis, acidogenesis, and methanogenesis can be individually dominant at separate periods within the test duration, with methanogens active mainly at the later stages of the tests. This is seldom the case in continuously fed commercial scale plants where all relevant microbial activities tend to take place simultaneously.

c. Microbial acclimatization that normally occurs and gradually improves with time in full-scale plants and can change the biodegradation rates does not occur to the same magnitude in BMP tests.

It is therefore necessary to complement BMP tests with pilot studies before considering full-scale application.

5

Posttreatment, Reuse, and Management of Co-Products

5.1 Biogas

5.1.1 Biogas Utilization

Table 1.1 (see Chapter 1) shows a typical composition of biogas from a good functioning anaerobic system. Under standard temperature and pressure (20°C and 1 atm), methane is a colorless and odorless gas and has a heating or calorific value of 35,000 kJ/m³. Carbon dioxide has no calorific value since it is non-combustible. Hence, the calorific value of biogas with 65% methane content is about 23,000 kJ/m³. Natural gas, on the other hand, has a much higher calorific value because of its lower carbon dioxide content. It also contains other hydrocarbons other than methane. Thus, by reducing the proportion of carbon dioxide in the biogas, the biogas caloric value can become comparable to that of natural gas, and both gases can be used for similar applications.

The primary use of biogas is as fuel, and may be used directly in a gas boiler for heating or burned in a gas engine to produce heat and electricity in a *combined heat and power* (CHP) process where about 70% of the energy contained in the biogas is converted to heat and the rest to electricity. The heat can be used to maintain the anaerobic system at mesophilic or thermophilic temperature, in supplementing the surrounding community heating requirements, particularly in temperate climatic regions, and for industrial uses. Biogas *upgrading* makes it suitable for other uses, such as injection into the district gas supply network for domestic, industrial, and vehicular uses. Upgrading to pure methane followed by liquefaction produces *biomethane* which creates an option for the biogas to be used as transport fuel for vehicles. More information on options for biogas use can be obtained elsewhere (German Solar Energy Society [DGS] and Ecofys 2005; IEA Bioenergy 2006, 2009, 2014; Polprasert 2007).

5.1.2 Biogas Treatment

Table 1.1 shows that biogas can contain trace concentrations of other gases such as oxygen, nitrogen, hydrogen, hydrogen sulfide, and ammonia. Impurities such as particulates and siloxanes may also be present. In practice, only CH_4, CO_2, and H_2S proportions are routinely monitored in a typical plant. Biogas treatment often involves the removal of moisture, hydrogen sulfide, and siloxane. Treatment for carbon dioxide reduction is referred to as *biogas upgrading*.

5.1.2.1 Moisture and Particulates Reduction

Moisture removal occurs naturally in the biogas collection piping system through condensation caused by the cooling effects of reduced temperature. Adequate pipeline design can provide sediment traps to capture both liquid and particulate matter produced by the condensation. These traps are normally located at the lower levels of the digester installation, and are emptied periodically. Longer piping networks may be required for more effective and durable performance. Figure 5.1 shows features of a typical pipeline design that can bring about moisture reduction. Where very dry gas is required, gas chilling may be necessary to improve condensation.

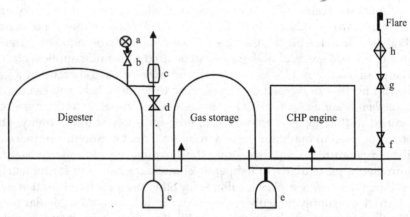

Key:

a. Air dosing pump
b. Back-pressure valve
c. Safety valve
d. Check valve

e. Condensing trap for moisture removal
f. Check valve
g. Self-closing valve
h. Flame trap

FIGURE 5.1
Schematic configuration of a typical biogas handling system. (Adapted from German Solar Energy Society (DGS) and Ecofys 2005).

5.1.2.2 Biogas Upgrading

Biogas upgrading becomes necessary where up to 99% methane content is required in order to increase its calorific value for greater efficiency in CHP engines, for use as transport fuel or for direct injection into municipal gas supply network. Biogas upgrading is achieved by reducing its carbon dioxide content. Carbon dioxide reduction can be carried using the following techniques:

a. Adsorption or dry scrubbing
b. Absorption or wet scrubbing using water and other reagents
c. Permeation (i.e., membrane separation)
d. Cryogenic upgrading

Wet scrubbing using water is the simplest method of reducing carbon dioxide contained in biogas. This method requires large quantities of water, which have to be replaced upon saturation. Other possible reagents include NaOH and $Ca(OH)_2$, and the reactions are shown in Equations 5.1–5.3.

With sodium hydroxide

$$2NaOH + CO_2 \rightarrow Na_2CO_3 + H_2O \tag{5.1}$$

$$Na_2CO_3 + CO_2 + H_2O \leftrightarrow 2NaHCO_3\downarrow. \tag{5.2}$$

With lime

$$Ca(OH)_2 + CO_2 \rightarrow CaCO_3\downarrow + H_2O. \tag{5.3}$$

Details of other upgrading techniques can be found elsewhere, e.g., IEA Bioenergy (2009).

5.1.2.3 Hydrogen Sulfide Removal

The concentration of hydrogen sulfide in biogas is dependent on the composition of the original wastewater or feedstock, and the system pH, as discussed in Chapter 1. Above 1% can cause significant metal corrosion that can damage engines and pipeworks. Some of the techniques for its reduction are discussed in the following section.

5.1.2.3.1 H_2S Removal inside the Digester

With oxygen (air or pure oxygen injection)

This is carried out by injecting 2%–4% volume of air into the headspace, and the chemical reaction that takes place is as shown in Equation 5.4. Too much air can lead to the production of sulfuric acid (Equation 5.5), which should be avoided. Care must be taken in mixing air and biogas to avoid explosion.

$$H_2S + 0.5O_2 \rightarrow S + H_2O \tag{5.4}$$

$$H_2S + 2O_2 \rightarrow H_2SO_4. \tag{5.5}$$

With metallic salts

H_2S can be removed through the formation of precipitates with iron or zinc salts. The salts combine with sulfide to form insoluble metallic salts, e.g., FeS, Fe_2S_3, F_3S_4, and FeS_2. The metallic salts can be mixed with raw feedstock and fed directly into the digester. A mixture of ferric and ferrous salts (Equation 5.6) is very effective.

$$Fe^{2+} + 2Fe^{3+} + 4H_2S \rightarrow Fe_3S_4 + 8H^+. \tag{5.6}$$

5.1.2.3.2 H_2S Removal by Absorption (or Wet Scrubbing)

H_2S is removed through dissolution in solvents such as water, chemical solutions containing oxidizing or reducing agents, and alkaline or acidic solutions. Examples are shown in Equations 5.7–5.11.

With sodium hydroxide

The chemical reaction is the same as in Equation 5.1. The Na_2CO_3 formed can be used to remove hydrogen sulfide (Equation 5.7) if longer retention times are provided.

$$H_2S + Na_2CO_3 + CO_2 + H_2O \rightarrow NaHS + NaHCO_3 \downarrow. \tag{5.7}$$

With hydrogen peroxide

Excess H_2O_2 decomposes to yield water and oxygen, which increases the dissolved oxygen content of the water, as shown in Equation 5.8.

$$H_2S + H_2O_2 \rightarrow S + 2H_2O. \tag{5.8}$$

With potassium permanganate

$$\text{In acidic conditions:} \quad 3H_2S + 2KMnO_4 \rightarrow 3S + 2H_2O + 2KOH + 8MnO_2 \tag{5.9}$$

$$\text{In alkaline conditions:} \quad 3H_2S + 8KMnO_4 \rightarrow 3K_2SO_4 + 2H_2O + 2KOH + 8MnO_2. \tag{5.10}$$

With chlorine

$$H_2S + 4Cl_2 + 4H_2O \rightarrow H_2SO_4 + 8HCl. \tag{5.11}$$

Others

a. Oxidation of hydrogen sulfide with sodium hypochlorite (normally used in conjunction with sodium hydroxide).

b. Ozone, dissolved in recirculating water, is an effective scrubber. Ozone has the advantages of being able to be generated on site and can minimize the build-up of reaction products in the scrubbing liquor.

5.1.2.3.3 H_2S Removal by Adsorption (Dry Scrubbing)

Adsorption is a process whereby molecules adhere to a surface with which they come in contact. To be effective, the *adsorbent* must possess relatively high surface area per unit volume or unit weight. Adsorption is carried out by passing the biogas stream through an adsorbent where the hydrogen sulfide is removed by attachment on its surface. For example, using iron filings (or ferric oxide) mixed with wood shavings, an adsorption reaction is shown in Equation 5.12.

$$Fe_2O_3 + 3H_2S \rightarrow Fe_2S_3 + 3H_2O. \tag{5.12}$$

Fe_2O_3 can be regenerated by exposing or heating Fe_2S_3 in air as shown in Equation 5.13

$$Fe_2S_3 + 3O_2 \rightarrow Fe_2O_3 + 3S_2. \tag{5.13}$$

Other adsorbents that can be used to remove hydrogen sulfide include the following:

- carbon (activated carbon, caustic impregnated carbon, charcoal, etc.)
- activated alumina
- activated bauxite
- aluminosilicate
- silica gel
- activated alumina impregnated with potassium permanganate

5.1.2.4 Simultaneous Removal of CO_2 and H_2S

Model examples of an application of wet and dry scrubbing techniques for the removal of carbon dioxide and hydrogen sulphide are shown in Figure 5.2.

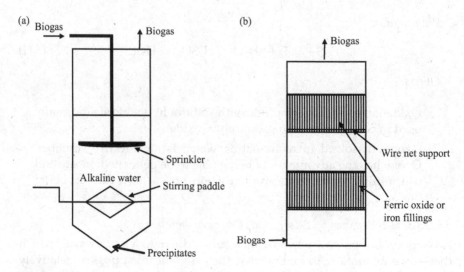

FIGURE 5.2
Model of (a) CO_2 and H_2S wet scrubber, and (b) H_2S dry scrubber. (Adapted from Polprasert 2007.)

5.1.2.5 Siloxanes Occurrence and Removal

Siloxanes are chemical compounds containing silicon, alkanes, and oxygen. They are widely used in a variety of industrial processes such as detergents, medical products and devices, shampoos, cosmetics, toothpastes, paper coatings, and textiles. Some of these siloxanes can end up in municipal wastewater. However, they do not generally breakdown in municipal wastewater treatment plants, so they end up as a component of biosolids. During anaerobic digestion, they can volatilize to trace quantities in the biogas. Silicone-based anti-foaming agents, used for controlling foaming in anaerobic reactors, can also augment the siloxanes content of the biogas. During combustion of biogas containing siloxanes, siloxanes are converted to silicon dioxide particles, similar to sand. These particles can cause abrasion of surfaces within the combustion chamber of CHP engines and sometimes cause significant mechanical damages. Siloxanes can be removed in combination with carbon dioxide and hydrogen sulfide using some of the aforementioned adsorption techniques. Lowering biogas temperature and pressure can also allow the compound to precipitate out into a liquid, and then settle out in condensation traps, as illustrated in Figure 5.1.

5.1.3 Health and Safety Considerations

The potential dangers associated with the operation of anaerobic treatment plants from feedstock and wastewater collection, system operation and to the management of co-/by-products are numerous. Operatives can be exposed to dangerous process gases from these activities.

A mixture of methane gas and air is combustible, and the flame is self-propagating in the proportions present in biogas.

Hydrogen sulfide gas is toxic to humans and animals. If inhaled at high concentrations, may lead to collapse, inability to breathe, and death within minutes. World Health Organization (WHO) guidelines on the dose effect of hydrogen sulfide are summarized in Table 5.1.

Hydrogen sulfide can also cause skin discoloration, pain, itching, skin redness and local frostbite, and eye irritation. It is very inflammable and can cause corrosion of metallic objects.

Ammonia gas production is also associated with anaerobic waste processing (see section 1.2.3.8.2 in Chapter 1). Exposure to ammonia gas can be fatal.

Accidents, such as explosion and asphyxiation by biogas or oxygen deficiency, and physical injuries caused by slippery surface, fall or dizziness by biogas inhalation, must be prevented through plant and operational design, and by adopting safe working practices which include the use of appropriate safety, devices, equipment, and clothing.

Some of the safety measures that can be addressed through plant design include the following:

- Large tank openings to ensure sufficient ventilation
- Use of safety valves at appropriate locations to prevent leakages, as shown for example, in Figure 5.1
- Valves and safety switches must be clearly labeled, easily identified and accessible
- Use of corrosion-resistant piping

TABLE 5.1

Dose–Effect Relationships for H_2S

H_2S Concentration		
mg/m^3	ppm	Effect
1,400–2,800	1,000–2,000	Immediate collapse with paralysis of respiration
750–1,400	530–1,000	Strong central nervous stimulation, hypopnea followed by respiratory arrest
450–750	320–530	Pulmonary edema with risk of death
210–350	150–250	Loss of olfactory sense
70–140	50–100	Serious eye damage
15–30	10–20	Threshold for eye irritation

Source: WHO (2000).

- Adequate ventilation to provide sufficient air movement in all operational areas, including treated and untreated wastes and wastewater collection, processing and storage areas, and biogas processing and utilization facilities, such as burner housing and CHP engines
- Ensuring safe distances between where biogas is produced, where it is re-used, and other non-operational areas such as the administrative buildings.

Ideally, operatives must undergo regular health and safety trainings and refresher courses.

5.2 Liquid Effluents

Depending on the sources of the original wastes or wastewaters and the management options for treated effluents, posttreatment may be required. As explained in Chapters 1 and 2, anaerobic wastewater treatment can only remove up to 80%–90% COD, and the rest is contained in the treated effluent. For wastes rich in nitrogenous compounds such as those from agricultural, food, and drinks processing industries and supernatants from biosolids digestion and dewatering operations, the treated effluent can contain elevated levels of solubilized ammonia and ammonium compounds. During anaerobic treatment, organic nitrogenous compounds are reduced to ammonium and used for cell growth, and the rest remain in the treated effluent. Posttreatment or *"effluent polishing"* before discharge may therefore be necessary in order to remove the undegraded organic matter only or with the ammonium depending on the environmental quality requirements of the receiving waterbody or other management options. Effluent polishing involves aerobic *carbonaceous oxidation* to reduce any remaining biodegradable organic compounds or combined with *nitrogenous oxidation* or *nitrification* to convert ammonia to nitrates. Where the effluent is to be discharged into surface water, which also serve as source of drinking water supply for downstream users or which is susceptible to excessive growth of algae or *eutrophication; denitrification* may be required to reduce the total nitrogen concentration. Aerobic carbonaceous oxidation, nitrification and denitrification processes have been covered in Chapter 1. Figure 5.3 shows examples of process configurations that can be combined with anaerobic treatment to bring about effluent polishing. Other considerations can be found elsewhere (e.g., Akunna and Bartie 2014).

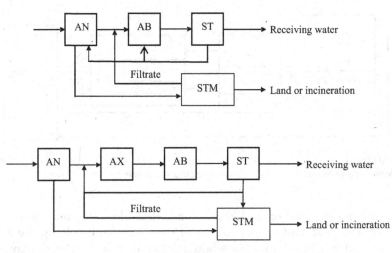

AB: Aerobic system
AN: Anaerobic system
AX: Anoxic system

ST: Sedimentation tank
STM: Sludge treatment and management

FIGURE 5.3

Typical flow diagrams for (a) anaerobic–aerobic treatment for carbonaceous pollutant reduction, and (b) Anaerobic–aerobic treatment for carbonaceous and nitrogenous pollutant reduction.

5.3 Digestate Management and Disposal

5.3.1 Characteristics and Management Options

Anaerobic digestion of solids or semi solids produces a semi-solid co-product, referred to as digestate. The solid content of the digestate will depend on the level of biodegradability of the solids in the original feedstock, and the effectiveness of any pretreatment that the feedstock may have been subjected to prior digestion. Only the readily biodegradable solids will normally breakdown during anaerobic digestion plant operation, leaving the non-readily biodegradable and inert solids in the digestate. Figure 5.4 shows a typical material flux in a biosolids anaerobic digestion system. The digestate volume is dependent on the moisture content of the original feedstock. The moisture content of the digestate is always greater than its amount in the original feedstock due to the breakdown of some of the biodegradable solids.

Consideration must therefore be given to potential end uses of the digestate and whether additional treatment such as dewatering for volume reduction or aerobic digestion (or composting) for further breakdown or *stabilization*

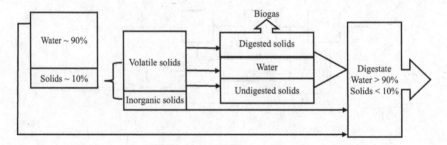

FIGURE 5.4
Typical material flux in a biosolids anaerobic digestion system (not to scale).

of any undegraded or undigested biodegradable solids may be necessary. Chemical treatment or *conditioning* of the digestate involving the addition of coagulants to enhance solid-liquid separation or *dewatering* may also be considered. Figure 5.5 summaries some of the options for digestate management. Table 2.1 (see Chapter 2) lists some of the coagulants that can be used for this purpose.

Digestates from anaerobic digested biosolids, agricultural, and food and drinks residues have generally been found suitable for land or soil application, as organic fertilizer, preferably following posttreatment operations notably dewatering and aerobic digestion (Abdullahi 2008; Abdullahi et al. 2008). The latter contributes to further stabilization and *pasteurization* of the digestate. Digestate pasteurization or disinfection for pathogen reduction before application on land used for agricultural purposes is mandatory in some countries.

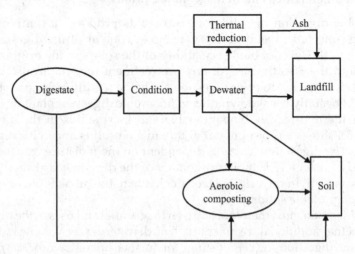

FIGURE 5.5
Options for digestate management and reuse.

5.3.2 Aerobic Composting

Depending on the characteristics of the original feedstock the major challenges in direct land disposal or application of digestate include high moisture content, low pH and odor nuisance. The digestate may also contain compounds that are resistant to anaerobic biodegradation (e.g., lignin) and high ammonia; some of these compounds could induce phytoxicity in the soil (Abdullahi 2008). Anaerobic digestate may therefore require aerobic posttreatment to render it more suitable for soil application. Aerobic composting or digestion can oxidize ammonium to its more stable form of nitrate, breakdown organic compounds that are more amenable to aerobic than to anaerobic biodegradation, and can reduce odor nuisance by degrading odorous organic compounds. Figure 5.6 shows a conceptual integrated system for digestate management and reuse.

Aerobic composting processes have been covered in Chapter 1. Operationally, it involves aerating the digestate with or without amendments (such sawdust and other available woody materials) with air or pure oxygen, in preferably closed vessels (this necessary in order to control odor nuisance). The biochemical reactions that occur under these conditions are shown in Equations 1.6. When the amount of biodegradable organic matter in the digestate is low, and aeration is continued, the produced microbial cells or new biomass begin to breakdown, releasing nutrients for use by the more viable bacteria to continue to survive. As substrate starvation and aeration continue, solids reduction also continues. This type of solids breakdown is called *endogenous respiration* or *cell lysis* and is represented by Equation 1.8. With continued aeration, nitrification will occur as indicated by Equation 1.9. Based on Equation 1.9, the theoretically oxygen demand for composting of biosolids can be estimated to be $1.42\,kgO_2/kgVS$.

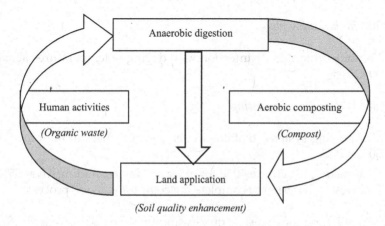

FIGURE 5.6
Integrated sustainable management of biodegradable organic solid waste (Abdullahi 2008).

5.3.3 Disinfection

Disinfection reduces the level of enteric pathogenic organisms in the digestate. There are regulations governing the level of pasteurization for biosolids
applied on land for agricultural purposes in many countries. Typical disinfection techniques include (Stenile 1993; UK Department for Environment
Food & Rural Affairs 2017):

Integrated in anaerobic digestion

- Pre-pasteurization. Minimum 30 min at 70°C or 4 h at 55°C followed
  by mesophilic digestion
- Mesophilic anaerobic digestion. Retention time of at least 30 days
- At least 12 days of primary digestion at 35°C ± 3°C or at least 20 days
  at 25°C ± 3°C followed by a secondary stage of at least 14 days
- Aerobic digestion (or composting) followed by thermophilic anaero-
  bic digestion
- Thermophilic anaerobic digestion
- Psychrophilic anaerobic digestion or long storage of untreated bio-
  solids for a minimum period of 3 months

Integrated in the dewatering phase

- Conditioning with lime, at pH 12 for a minimum of 2 h.
- Conditioning with lime or other coagulants followed by dewatering
  and storage for 3 months. If biosolids has been subject to primary
  mesophilic anaerobic digestion, storage can be for 14 days.
- Thermal conditioning at temperatures above 60°C

Integrated in the drying phase

- Combined thermal disinfection with drying at temperatures above
  60°C

Integrated with aerobic composting

- Composting systems that ensure temperature rises to the range of
  50°C–70°C.
- Composting to achieve 40°C for at least 5 days including 4 h at 55°C
  followed by a period to complete the compost reaction process

Further reading on the digestate quality and reuse regulations can be found
elsewhere, e.g., IEA Bioenergy (2010).

6

Applications in Warm Climates and Developing Countries

6.1 Characteristics of Warm Climatic Conditions

With regards to general waste management, tropical weather may bring about some of the following conditions and considerations:

- Low external energy requirements for biological processes due to high ambient temperatures
- Risk of losses of biodegradable organic carbon through uncontrolled decomposition during waste collection, processing, and storage due to high ambient temperatures
- Higher risk of odor and flies nuisance in waste management facilities due to accelerated waste decomposition in collection and storage facilities
- Greater use of treatment technologies that require low energy and large spaces
- Greater occurrence of natural by the sun's ultraviolet radiation
- Greater need for water and nutrients recovery and reuse for agriculture due to longer periods of dry weather
- Greater risk of water pollution caused by low dilution rates in receiving waters, particularly during periods of dry weather.

The above attributes impact on waste management practices and regulations in these regions.

6.2 Characteristics of Developing Countries

Developing countries are usually associated with some of the following attributes:

- High population growth and density in urban and peri-urban areas
- Water supply sources are highly venerable to anthropogenic sources of pollution
- High rates of freshwater losses by evaporation
- Limited access to affordable artificial fertilizers for soil improvement
- Poorly funded waste and wastewater collection in urban and peri-urban areas
- Stigmatizing cultural customs and norms with regard to handling and working with municipal and human wastes
- Weak governance structures for environmental management, regulations and enforcement regimes
- Limited skilled manpower
- High rate of deforestation due to use of trees for firewood
- Poor data acquisition and management
- Insufficient and unreliable power supply
- Women and children play leading role in water, waste, and sanitation issues.

Technology and energy remain the driving factors of any economy. The costs at which these are produced and delivered determine the economic viability and competitiveness of any country. Low energy waste treatment technologies are usually low cost although they occupy more space than more effective but capital-intensive high-energy demand technologies. Considering the ever-growing global energy demand, there is need for developing low energy systems and utilizing renewable energy sources such as biogas to complement the current high global reliance on fossil energy sources. Biogas technology can contribute toward lowering demand for fossil fuels, fuel wood, and mineral fertilizers, even as it ensures a cleaner environment through improved sanitation and waste management. Water and fuel wood are vital natural resources, and their collection and use in rural households is often the responsibility of women and girls. Biogas technology, represents a reliable and cheap source of locally produced energy and can contribute in improving the standard of living of poor rural communities.

6.3 Waste and Wastewater Characteristics

Waste generation rate is greater in developed than in developing countries, with per capita estimates varying from 1.43–2.08 and 0.3–1.44 kg/day, respectively; with the organic fraction accounting for 55%–60% in developing countries, more than twice the value in some developed countries (Friedrich and Trois 2011; Mucyo 2013; Troschinetz and Mihelcic 2009). This is due to a greater proportion of food wastes produced by the former. Food wastes are generally wet, with some components containing more than 70% moisture. Thus, the quality of domestic wastes produced in most developing countries is suitable for anaerobic digestion. However, uncontrolled decomposition during collection, transporting, and storage, aided by high ambient temperatures, can significantly reduce their biogas yield. Furthermore, due to the relatively low water usage in these countries (per capita consumption of 50–100 L/day in some water stressed regions compared to 180–220 L/day in Western Europe) and high water loss through evaporation, domestic wastewater (which is usually collected in septic tanks), has much higher concentrations of biodegradable organic matter and nutrients than those produced in developed countries. Septic tank effluent is, therefore, a suitable feedstock for the production of biogas and organic fertilizer.

6.4 Wastewater Treatment

6.4.1 Large-Scale Systems

The anaerobic upflow filter (UF) and upflow anaerobic sludge blanket (UASB), described in Chapter 2 are two of the most common systems for large-scale municipal and industrial applications. UF system requires prior solids separation in sedimentation or septic tanks, to reduce clogging of the support media and consequent flow short-circuiting. Locally obtainable media, such as plastics, stones, etc., can be used for microbial support. UASB systems can tolerate high organic solids loading; hence, they do not normally require a pre-sedimentation step. Both systems, however, require prior screening out of large and floating objects, and removal of inert materials, such as sand.

Detailed information on design and operation UF and UASB systems can be found elsewhere (e.g., Malina and Pohland 1992; Metcalf and Eddy 2014; Polprasert 2007; von Sperling and de Lemos Chernicharo 2005). Depending on ambient temperature and its level of seasonal fluctuations, anaerobic treatment can be carried out without extra heating during high temperature seasons, and where practicable and affordable, extra heating provided during low temperature periods. Posttreatment may be required if the effluents are to be discharged into waterbodies.

6.4.2 Micro-Scale Systems

These are portable or package anaerobic digesters for small-scale applications, for example, individual households, hotels, etc. They are usually unheated and unmixed. Feeding mechanisms are designed to encourage plug flow within the system and prevent excessive flow short-circuiting. Large micro systems are commonly used in prisons, schools, camping sites, etc. in parts of Asia and Sub-Saharan Africa. Common feedstocks include grasses, weeds, crop harvest residues, and kitchen, human, and animal wastes. Micro-scale systems normally consist of feeding hopper, digester, and gas collection and storage units. They are designed to operate at ambient temperatures, and sometimes buried underground to ensure good insulation and reduce temperature fluctuations. Underground installations also enable gravity feeding. Biogas yields are variable with high values occurring during high temperature seasons and when feedstocks are abundant. Figure 6.1 shows two examples of such systems. More information can be found in Polprasert (2007).

Biogas produced by these systems is mainly used for cooking and household lighting. Affordable small engines that run on duel diesel–biogas engines now exist, and these have enabled biogas to be used for other purposes such as refrigeration, electricity generation, and running irrigation pumps. Flexible biogas storage bags made of plastic materials are also available. The biogas quality can be enhanced by wet scrubbing (see Chapter 5) typically with water, which reduces its carbon dioxide and hydrogen sulfide content. The water has to be replaced regularly due to low pH caused by the dissolved gases.

The digestate can contain pathogens (the amount usually high where human and animal wastes form part of the feedstock), and thus requires careful handling and disposal. Open drying beds, an example shown in Figure 6.2, can reduce moisture content and achieve natural ultra violet pasteurization. The drying bed must be properly designed to avoid interference by rainfall runoff, and to prevent the filtrate escaping and contaminating surrounding environment or infiltration into the groundwater. The digestate is often used as organic fertilizer or simply applied to land used for growing pastures for grazing animals. It can also be used as fish food in aquaculture.

6.4.3 Waste Stabilization Ponds

Waste stabilization ponds often form part of a decentralized domestic wastewater treatment system in developed and developing countries alike, where they are main treatment and/or posttreatment. In temperate regions, their use is often restricted to tertiary and final polishing following aerobic or anaerobic treatment. While in tropical climates, they can represent both

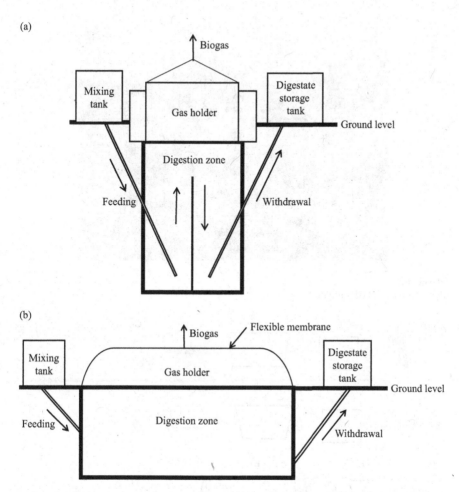

FIGURE 6.1
Schematic diagrams of some micro-scale plug flow anaerobic digestion systems equipped with (a) floating gas holder and (b) flexible membrane gas collection and storage bag.

main (i.e., solids sedimentation and biological treatment) and posttreatment. Waste stabilization ponds are effective, low cost technologies; however, they require large space, and hence, only suitable for rural populations where land is readily available.

Ponds can be designed to receive wastewater directly from surrounding populations, and septic tanks effluents from afar. The transport of septic tank effluents can sometimes pose technical challenges due to odor and occasional breakdown of vehicles. The risks in transfer of septic effluents must be weighed against the potential benefits of the pond location, such as less proximity to population, land costs, etc. Figure 6.3 shows system configurations of some applications of waste stabilization ponds.

FIGURE 6.2
Open-air biosolids drying bed.

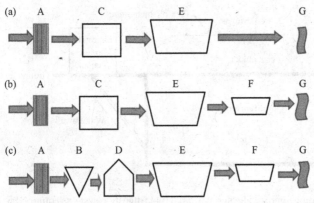

Where,
A: Screening to removal gross and floating solids
B: Grit or sand removal
C: Septic tank or covered anaerobic lagoon or pond
D. Anaerobic covered lagoon or pond
E: Facultative pond. Alternatively, Aerobic pond or oxidation ditch followed by sedimentation basin
F: Maturation pond or land treatment
G: Receiving waterbody or effluent reuse, such as irrigation.

FIGURE 6.3
Typical flow sheets for the application of waste stabilization ponds in wastewater treatment.

Figures 6.3a and b are examples of system configurations for rural and decentralized applications, while Figure 6.3c is suitable for urban or peri-urban applications. Organic solids reduction takes place in the septic tanks, where low- and high-level hydrolytic activities can occur depending on the tank temperature. At temperatures greater than 15°C, significant levels of solids biodegradation can be achieved, resulting in less frequent desludging. Septic tanks and ponds are usually designed for desludging intervals of 3–5 years, or when the units are half-full. At temperatures less than 15°C, the rate of solids biodegradation is generally very low and the tanks act merely as storage and settling basins, and desludging becomes more frequent.

Septic tank effluents contain low concentrations of volatile suspended solids and variable levels of dissolved organic compounds, which can be removed by a combination of aerobic and anaerobic treatment processes that take place in facultative ponds. Figure 6.4 is a schematic diagram of a facultative pond, showing the major physical and biochemical activities that can take place therein.

Maturation ponds, where needed, follow facultative ponds. These are aerobic ponds where effluent polishing and disinfection can take place. This is a vital stage if the treated effluent is to be discharged into receiving water bodies used for recreational and sporting activities, bathing, or shellfish farming. Disinfection in maturation ponds occurs through microbial predation and lysis caused by starvation, and the sun's ultraviolet radiation. The cysts and ova of intestinal parasites settle at the bottom of the pond where they eventually die.

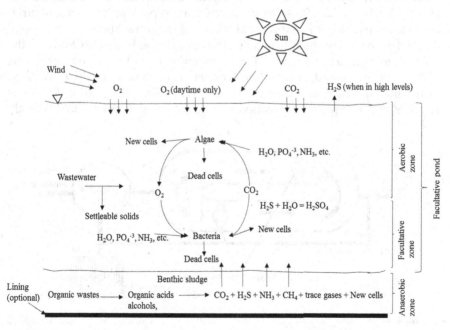

FIGURE 6.4
Waste stabilization pond processes. (Adapted from Crites and Tchobanoglous 1998.)

Aerobic ponds can be used instead of facultative ponds to improve the rate of biodegradation and by so doing reduce its footprint. Aerobic ponds are completely mixed and aerated using mechanical systems, such as motor-driven surface aerators, to bring about aerobic conditions to all parts of the pond and to increase the microorganism–waste contact. The resulting increase in organic biodegradation rate enables the pond to be smaller than the facultative pond. Aerobic ponds are usually followed by sedimentation basins to settle out the produced biosolids before effluent discharge. Desludging frequency of a sedimentation basin is dependent on the organic loading or F/M ratio (Equation 2.6 in Chapter 2), and can vary from 9 months to a few years. The waste biosolids can be processed for reuse as organic fertilizer. With proper design, up 85%–90% BOD removal can be achieved in aerobic ponds. The sedimentation tanks are usually followed by maturation ponds if treated effluent disinfection is required.

Oxidation ditches can be used in place of aerobic ponds for application in high population areas. They are similar to the aerobic ponds, except that they involve recycling produced biosolids (or sludge) and forward movement of the wastewater within the pond or ditch. Figure 6.5 shows a schematic diagram of an oxidation ditch and picture of a real system in Figure 6.6. The forward movement, aeration, and sludge recycling are provided by rotating cylindrical "brushes," which drive the waste water forward and simultaneously provide mixing and turbulence that enhance natural oxygen transfer from the atmosphere and ensure that aerobic conditions are maintained at every part of the ditch. Biosolids co-product are separated from treated effluent in a separate sedimentation tank. Some of the settled biosolids are regularly returned to the ditch to achieve the operational F/M ratio. Some of the settled biosolids are also periodically removed from the system, as waste, and can be processed directly for production of organic fertilizer or used in the first instance as feedstock for anaerobic digestion. Up to 90% BOD removal and complete nitrification can be achieved in an oxidation ditch.

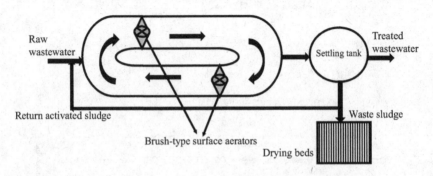

FIGURE 6.5
Schematic diagram of an oxidation ditch.

FIGURE 6.6
A picture of an active oxidation ditch.

As in aerobic pond treatment, further treatment in maturation ponds may be necessary if disinfection is required.

Equations 2.1, 2.2 and 2.6 (in Chapter 2) can be used for the design and operation of aerobic ponds and oxidation ditches.

Off-gas from the septic tanks and anaerobic ponds is composed mainly of carbon dioxide and hydrogen sulfide, and therefore constitute odor nuisance if it is not collected and disposed properly. Collection is normally via long vertical pipes and high enough to ensure sufficient atmospheric dilution.

A major benefit of including anaerobic ponds and septic tanks in wastewater treatment systems in tropical climate, even when the organic content of the wastewater is too low for viable biogas recovery and utilization, is their positive contribution in the initial reduction of organic content, which can lead to the use of the relatively low cost ponds systems for further and/or posttreatment.

6.5 Solid Wastes and Slurries Treatment

The most common global application of anaerobic treatment technology is in digestion of solid wastes and slurries. The uptake of low cost micro-scale systems in most developing countries is currently on increase, due probably to increased rural energy demand in recent years driven by the general increase in global rural standard of living in these countries boosted by affordable mobile electronic communication tools and personal computers. However, by comparison, the uptake for large-scale urban OFMSW digestion systems has been slow (Mucyo 2013; Mucyo and Akunna 2014) due partly caused by the high capital costs associated with large operations. Alternative community composting projects have also not been widespread due to low revenues from the compost sale (Couth and Trois 2012). Consequently, traditional cheaper disposal routes such as landfills and open dumps are still play a major role in waste management even as these routes present high public health risks and environmental costs.

7

Case Studies

7.1 Brewery Wastewater Treatment Using the Granular Bed Anaerobic Baffled Reactor (GRABBR)

This study was aimed at evaluating the anaerobic treatability of brewery wastewater and to evaluate the effectiveness of the granular bed anaerobic baffled reactor (GRABBR) system, developed by the author and his team (Akunna and Clark 2000; Baloch and Akunna 2003a, b; Baloch et al. 2007). The study was carried out at the premises of one of the biggest breweries in Scotland. The brewing processes and operations produced different wastewater streams which were separately collected. Two of these had relatively high COD levels and represented the majority of the cost of effluent discharges incurred by the company. For one of the streams, the majority of the COD was composed of organic solids, which originated from the yeast evaporator and centrifuge concentrate. The majority of COD in the other stream was soluble. Both streams varied significantly in both quality and quantity. These variations were minimized by collecting both in a *balancing tank*.

The treatment process was operated on-site as a continuous flow process. Figure 7.1 shows a schematic diagram of the treatment system. From the balancing tank, the effluent was pumped at a controlled rate into a proprietary sedimentation unit to remove rapid-settling solids. The overflow from the unit emptied into a *conditioning tank* where pH and micronutrients corrections where carried out before being fed to the GRABBR. The average characteristics of the wastewater fed to the GRABBR were typically COD 10,000 mg/L, TS 500 mg/L, and pH 8. Heating and temperature control was carried out by recycling a proportion of the treated effluent from the GRABBR to the conditioning tank via an electrical heater to raise its temperature to 35°C ± 1°C.

The GRABBR, of 10 m^3 effective capacity, consisted of six compartments. The seed inoculum was obtained from an upflow anaerobic sludge blanket (UASB) reactor treating paper mill wastewater. The wastewater traveled through the GRABBR compartments by gravity through 100 mm interconnecting pipes before overflowing from the system to drain or recycled to the conditioning tank.

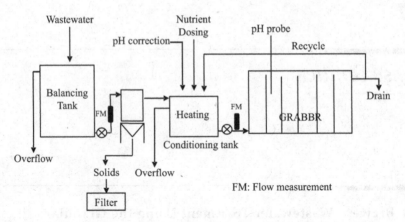

FIGURE 7.1
Process flowsheet of the GRABBR pilot plant.

Daily variations of influent and effluent COD over four typical weeks of operation following start-up is shown in Figure 7.2. The figure shows that even with the balancing tank, the COD levels of wastewater fed into the GRABBR were still highly variable but the system performed well.

Figure 7.3 shows the COD and suspended solids profiles within the system compartments when operating at OLR of 13 kg COD/$m^3 \cdot$day. The overall process efficiency generated at this loading was over 90% reduction of COD. The slight increase in level of suspended solids observed in the treated

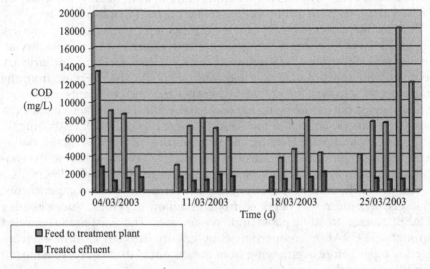

FIGURE 7.2
Daily variations of the GRABBR influent and effluent COD concentrations.

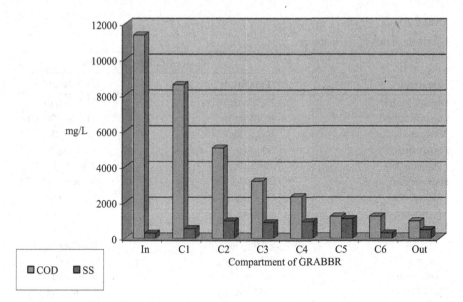

FIGURE 7.3
In-compartment and final effluent COD and SS profiles at OLR of 13 kg COD/m$^3 \cdot$day.

effluent was primarily due to the production of a dispersed biomass in the early compartments of the GRABBR.

Figure 7.3 also shows that only the first four compartments of the GRABBR played a major role in COD reduction; hey were subjected to an equivalent OLR of 20 kg COD/m$^3 \cdot$day and achieved 80% COD removal efficiency. This implies that the entire system still had capacity for higher OLR.

The results of this case study showed the following:

- Brewery wastewater is amenable to anaerobic treatment.
- The GRABBR system is an effective plug-flow phase separation technology, which can *accommodate high fluctuations* in wastewater characteristics.
- Flow balancing is an important pretreatment in anaerobic industrial wastewater treatment.

7.2 Seaweed Anaerobic Digestion

Marine macroalgae or seaweed is regarded as potentially an important source of feedstock for the production of bioenergy through anaerobic digestion. However, seaweed salinity has been reported as an impediment to effective digestion. The aim of this study was to determine the effects of

TABLE 7.1

Characteristics of *A. nodosum*

| Parameter | Concentration (Dry Matter) | |
	This Study	Literature[a,b,c]
Organic matter	75.52% m/m	74.0% m/m
Organic carbon	43.9% m/m	34.7–36.1% m/m
Nitrogen (N)	16.2 g/kg	6.5–20 g/kg
Phosphorus (P)	3.0 g/kg	1–2 g/kg
Potassium (K)	24.4 g/kg	20–30 g/kg
Sodium (Na)	25.5 g/kg	30–50 g/kg
Calcium (Ca)	8.0 g/kg	10–30 g/kg
Magnesium (Mg)	5.8 g/kg	5–10 g/kg
Sulphur (S)	0.85 g/kg	25–35 g/kg
Copper (Cu)	9.0 mg/kg	6 mg/kg
Iron (Fe)	383.0 mg/kg	600 mg/kg
Manganese (Mn)	84.0 mg/kg	30 mg/kg
Zinc (Zn)	138.0 mg/kg	108 mg/kg
Cadmium (Cd)	0.07 mg/kg	NR
Chromium (Cr)	3.0 mg/kg	NR
Nickel (Ni)	6.0 mg/kg	3 mg/kg

NR, not reported.
[a] Bruton et al. (2009).
[b] Gevaert et al. (2008).
[c] Ross et al. (2008).

salinity and reactor configuration on the effectiveness of seaweed anaerobic digestion. The seaweed species, *Ascophyllum nodosum*, and two reactor configurations, single- and two-stage mesophilic reactor systems were used for the study. The chemical characteristics of dried and milled *A. nodosum* or feedstock used in this study are compared with those reported in the literature in Table 7.1. Due to unavailability of saline acclimatized anaerobic sludge, seed inoculum comprised of a mixture of decaying seaweed and anaerobically digested municipal wastewater sludge. Various quantities of the feedstock were mixed with freshwater in order to reduce its salt content and to obtain 2% and 5% total dry solids.

Data presented in Table 7.1 indicated that the composition of the seaweed used in the study is comparable with those reported in the literature. Notable is the relatively high amount of sodium content.

The single-stage and two-stage systems were operated for 94 and 78 days, respectively. The operational conditions are summarized in Table 7.2. Feeding was suspended between Day 37 and 57 (i.e., for 20 days), due to excessive VFA build up and low biogas production, and restored on Day 58 with a reduced OLR.

TABLE 7.2

Summary of Operating Conditions

Reactor Configuration	Operation Duration (day)	Operating Condition	Total Solids Concentration % (TS)	OLR kg TS/m³·day (kg VS/m³·day)	HRT (day)
Single-Stage	1–21 (21 days)	Start-up Semi-continuous operation	—	—	—
	22–37 (16 days)	Discontinued after 16 days due to poor performance	5% (50 g/L)	2.3 (1.74)	22
	38–57 (20 days)	No feeding	—	—	—
	58–92 (37 days)	Semi-continuous operation	2% (20 g/L)	1.1 (0.83)	18
Two-Stage	1–29 (29 days)	Start-up Semi-continuous operation			
	30–37 (8 days)	Discontinued after 8 days due to poor performance	5% (50 g/L)	2.3 (1.74)	22
	38–57 (20 days)	No feeding	—	—	—
	58–78 (21 days)	Semi-continuous operation	2% (20 g/L)	1.1 (0.83)	18

The overall performances of the reactors, during and after start-up, are shown in Figures 7.4 and 7.5. Start-up for the single-stage and two-stage systems lasted for 21 and 29 days, respectively, as shown in the figures. During this period, the single-stage system and the methanogenic reactor of the two-stage system were operated as single-stage systems with low OLR. The acidogenic reactor of the two-stage system was started on Day 30.

Figures 7.4 and 7.5 show that the two-stage system performed significantly better than the single-stage system for the same OLR, suggesting that the former provided more favorable conditions for methanogens to acclimatize to the substrate.

The specific methane yield for the single-stage and two-stage systems at steady state were in the range of 0.10–0.15 LCH_4/VS at standard temperature and pressure (i.e., 273.15 K and 1,013.25 kPa).

Results of the study show that the the seaweed species used in the study, A. *nodosum*, is amenable to anaerobic digestion for biogas production at high freshwater dilution and low OLR. The study also showed two-stage digestion to be more effective than a single stage digestion in ensuring process stability.

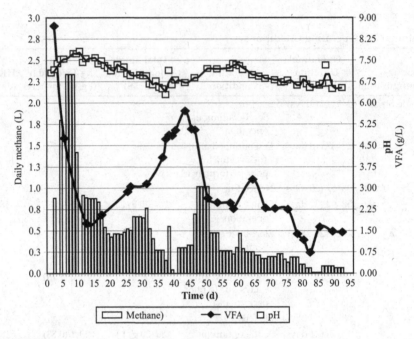

FIGURE 7.4
Single-stage system.

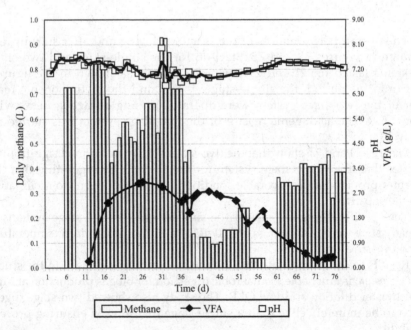

FIGURE 7.5
Methanogenesis reactor of the two-stage system.

7.3 Seaweed Anaerobic Co-Digestion

The study investigated factors affecting *co-digestion* of seaweed with terrestrial plant biomass, garden peas (*Pisum sativum*). The latter was obtained from a grocery store. A brown seaweed of species, *Laminaria digitata*, was

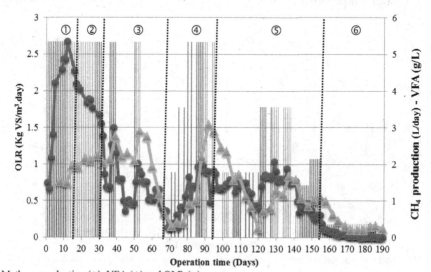

Methane production (●), VFA (▲) and OLR (■)

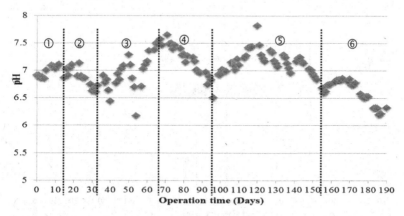

①. Green peas only: continuous feeding
②. Green peas + 2% (dry wt) seaweed: continuous feeding
③. Green peas + 2% (dry wt) seaweed: intermittent feeding
④. Green peas + 2% (dry wt) seaweed: intermittent feeding
⑤. Green peas only: intermittent feeding
⑥. Green peas + 2% (dry wt) seaweed: continuous feeding

FIGURE 7.6
Performance of high start-up OLR co-digestion over 190 days (Akunna and Hierholtzer 2016).

collected from Westhaven beach (56° 30′ N, 2° 42′ W) near Dundee, Scotland, UK in October 2010 and October 2011. After collection, the seaweed was washed with freshwater to remove debris, sand and excess seawater. Both feedstocks were oven-dried at approximately 75°C for 24 h and milled in an industrial blender to reduce particle size to a maximum of 1 mm and obtain homogenized feedstocks. The feedstocks were then stored in sealed containers at room temperature.

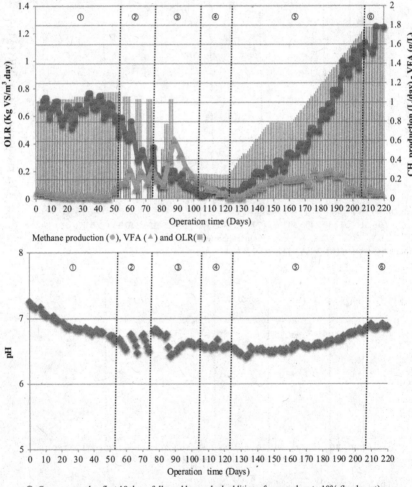

Methane production (●), VFA (▲) and OLR(■)

①. Green peas only , first 10 days, followed by gradual addition of seaweed up to 10% (by dry wt)
②. Green peas with gradual reduction in the proportion of seaweed to about 2% (by dry wt)
③. Green peas + about 2% (by dry wt) of seaweed: decrease in overall OLR
④. Green peas with gradual increase in the proportion of seaweed: constant low OLR
⑤. OLR increase and increase in the proportion of seaweed to up to 35% (by dry wt)
⑥. Green peas + 35% (dry wt) seaweed

FIGURE 7.7
Performance of low start-up OLR co-digestion over 220 days (Akunna and Hierholtzer 2016).

Varying weight ratios of each of the feedstock were mixed with 300 mL of tapwater before addition to a 5 L capacity continuously stirred-tank reactor (CSTR) reactor seeded with mesophilic anaerobically digested municipal wastewater sludge . The reactor was operated at an optimum mesophilic temperature (37°C ± 1°C) and fed once daily with a 17 day hydraulic retention time (HRT). Two distinct experiments were conducted, firstly with high start-up OLR. This experiment was discontinued due to process instability caused by excessive accumulation of VFA and resulting low pH, and the second experiment started with a reduced initial OLR. Results of both experiments are shown in Figures 7.6 and 7.7, respectively.

Results showed that when 100% of garden peas was used as feedstock at an initial OLR of 2.67 kg VS/m^3·day, steady state performance and process stability were established. However, when only 2% of the feedstock was replaced with the seaweed, with the total OLR maintained constant, methane production was disrupted, resulting in excessive VFA accumulation as shown in Figure 7.6. Process stability was difficult to achieve thereafter. When the experiment was repeated with a much lower initial OLR of green peas of 0.70 kg VS/m^3·day and 2% of the total OLR replaced with seaweed, greater process stability was observed than in the first experiment. This enabled a gradual increase in the proportion of seaweed to up to 35% of a total OLR of 1.25 kg VS/m^3·day.

The results of this study suggest that in addition to salinity, seaweed may contain other compounds that can adversely affect its anaerobic digestibility. These other compounds, even when present in trace concentrations (such as 2%), can disrupt the process, particularly the methanogenesis. Follow-up studies have shown that some types of phenolic compounds fall within this category of inhibitory compounds (Hierholtzer et al. 2013). Hence, seaweed anaerobic digestion requires long periods at low OLR to enable effective acclimatization of microorganisms to the inhibitory constituents. Where there are seasonal variations in the availability of seaweeds as feedstock for co-digestion applications, repeat acclimatization procedures may be required during each period of use (Akunna and Hierholtzer 2016).

Appendix A: Worked Examples on Anaerobic Wastewater Treatment

Question A1: Industrial Wastewater Management and Process Selection

Table A.1 shows average characteristics of wastewater streams from an alcohol beverage fermentation process. The local environment regulation requires any wastewater discharging into a nearby river not to exceed SS, BOD, and COD concentrations of 30, 20, and 125 mg/L, respectively, and pH 6–8. It is required to propose a treatment strategy for the wastewater streams for discharge into the river.

Solution A1

A wastewater strategy for the site and treatment options are shown in Figure A.1.

The balancing tanks will ensure flow of uniform characteristics to the treatment systems and to the river. Balancing Tank I should be insulated to reduce heat loses.

The Second Wash stream can be used for pH correction; hence, only the quantity necessary for pH correction should be added to Balancing Tank I and the rest diverted to Balancing Tank II. Where a significant amount of the alkaline wastewater (i.e., Second Wash) stream is added to Balance tank II, pH correction may be required before discharging in to the river.

The aerobic treatment stage will require a sedimentation tank to separate the biosolids co-products treated effluent. Excess biosolids from the aerobic treatment can be dewatered and recycled into the anaerobic reactor to boost biogas production. While waste biosolids from the anaerobic treatment can be dewatered and used as soil conditionner (see Figure 5.5, Chapter 5).

TABLE A.1

Characteristics of Wastewater Streams

Process	Nature of Effluent	Composition			
		COD (mg/L)	BOD (mg/L)	SS (mg/L)	pH
Pretreatment	Feedstock washing	120	10	20	8.0
Fermentation	Residual by-products	66,000	25,000	16,000	5.0
Cleaning of	First Wash	67,000	26,000	600	4.0
fermentation	Second Wash	200	50	30	11.0
vessels	Third Wash	85	18	20	7.0

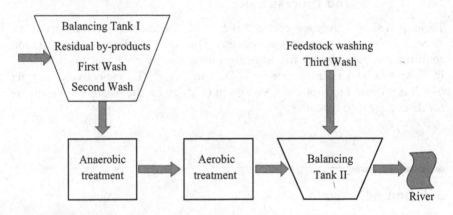

FIGURE A.1

Flow sheet of a possible treatment options.

Question A2: Process Selection, Reactor Sizing, and Biogas Yield Estimation

The average wastewater characteristics produced by a food processing plant are shown in Table A.2.

If anaerobic treatment is proposed as a treatment or pretreatment option, provide initial working plans and estimates for the following:

a. Pretreatment requirements

b. Reactor size

c. Biogas production

d. Sludge (or biosolids) production

TABLE A.2

Average Wastewater Characteristics

Parameter	Value
Flow rate, m³/day	500
COD, mg/L	11,000
BOD_U, mg/L	10,000
TKN, mg N/L	200
TSS, mg/L	2,000
VSS, mg/L	1,800
$SO_4^=$, mg/L	200
Alkalinity, mg/L $CaCO_3$	500
Temperature, °C	30
pH	4.8

Solution A2

a. Pretreatment

pH correction: Assume low pH is caused by fermentation of readily biodegradable organics contained in the wastewater. Provide balancing tanks to correct pH, using NaOH, KOH, or $Ca(OH)_2$. It is possible that some of the sulfate will be reduced to H_2S during the fermentation process, so the balancing tank should be covered and the produced biogas, which will be mainly CO_2 and H_2S, collected and safely disposed of. The balancing tank should be insulated to reduce heat loses.

Nutrient correction: COD/N ratio is about 55, which is within acceptable rage for anaerobic treatment (see Chapter 1).

Solids: These are assumed biodegradable since the wastewater originated from food processng activities.

Alkalinity: Satisfactory.

b. Tank sizing

With relatively low solids content, and a significant proportion of readily biodegradable organic matter, and possibility of pre-fermentation in the balancing tank, assume a 20 day hydraulic retention time (HRT) in a completely-mixed reactor without further heating in the first instance.

(*Note: BMP and pilot-scale studies must be carried out to establish the retention time that will meet the desired treatment objective*).

Using Equation 2.1 (Chapter 2)

Tank volume (without allowance for biogas storage) =

$$HRT \times Flow\ rate = 20\,day \times 500\,m^3/day = 10,000\,m^3$$

This provides 4 tanks, diameter 20 m, and height 10 m. These tanks will be covered with expandable membrane covers and equipped with facilities for biogas collection for utilization (see Figure 5.1, Chapter 5).

c. Methane production

Using data from Table 2.5 (Chapter 2), assume 80% BOD reduction (this will be determined from BMP and pilot studies). Note that the COD is almost equal to the ultimate BOD, meaning that most of the COD is biodegradable for this wastewater.

$$CH_4 \text{ production} = 0.35 \text{m}^3 \left[(0.8)(500 \text{m}^3/\text{day})(11,000 \times 10^{-3} \text{kg COD}/\text{m}^3) \right]$$

$$= 1,540 \text{m}^3/\text{day}$$

Net methane production will be lower if heating is carried out to bring the operating temperature to the optimum mesophilic value of approximately 37°C.

d. Sludge production and storage

Using data given in Table 2.5.

Total organic solids content of the sludge

$$= 0.06 \left[(0.8)(500 \text{m}^3/\text{day}) (11,000 \times 10^{-3} \text{kg COD}/\text{m}^3) \right]$$

$$= 264 \text{kg VS}/\text{day}$$

Assuming the sludge solids contains 90% VS and the rest inorganic salts.

Then, TS = 264/0.9 = 293 kg /day

Assuming the TS represents only 2% of the sludge, with moisture making up the rest

$$\text{Sludge production} = 14,650 \text{kg}/\text{day}$$

Assuming the sludge has a specific gravity of 1.02, volume = 14.4 m³/day

In practice, the sludge conditioning using coagulants (Table 2.2, Chapter 2) followed by dewatering can increase the solids content to over 30%. A 30% solids content is equivalent to less than 0.4 m³ of daily sludge production, i.e., a reduction in sludge storage facility of over 97%.

The above example is required for initial analysis of space and material requirements. Detailed design and operational data should always be obtained from laboratory and pilot studies before full-scale application.

Appendix B: Worked Examples on Anaerobic Digestion of Solid Wastes and Biosolids

Question B1: Agricultural Waste Anaerobic Digestion

Determine the monthly potential bioenergy production from anaerobic digestion of food processing wastes with the following characteristics:

$$\text{Feedstock production} = 1,000 \, \text{tonne/month}$$

$$\text{Moisture content}(MC) = 90\%$$

$$\text{Volatile solids}(VS) = 90\% \text{ of total dry solids}(TS)$$

$$\text{Biodegradable organic solids} = 80\% \text{ of VS}$$

$$\text{Minimum biogas yield} : 500 \, \text{m}^3/\text{tonne} \cdot VS$$

Assume that the minimum calorific value of biogas as 23,000 KJ/m³.

Solution B1

Total readily biodegradable dry organic solids (i.e., biodegradable VS)

$$= 1,000 \times (10\%) \times (90\%) \times (80\%) = 72 \, \text{tonne/month}$$

$$\text{Total biogas} = 500 \times 72 \, \text{m}^3 = 36,000 \, \text{m}^3/\text{month}$$

$$\text{Energy production} = 36,000 \times 23,000 \, \text{KJ/month} = 828 \times 10^6 \, \text{KJ/month}$$

Question B2: Industrial Waste Anaerobic Digestion

Estimate the total theoretical amount of biogas that could be produced from anaerobic digestion of organic waste having a chemical formula $C_{50}H_{100}O_{20}N_5$. Assume the reactor is operated at a retention time of 20 day, and only 80% of the organic matter will have biodegraded during this period.

Note:
Calorific value of biogas: 23,000 KJ/m³
Density of methane gas (CH_4): 0.7167 kg/m³
Density of carbon dioxide (CO_2): 1.9768 kg/m³
Atomic weights: C (12.0); H (1.0); N (14.0); O (16.0)

Solution B2

Using Equation 3.3 (Chapter 3)

$$C_aH_bO_cN_d + 0.25(4a-b-2c+3d)H_2O \rightarrow 0.125(4a+b-2c-3d)CH_4$$
$$+ 0.125(4a-b+2c+3d)CO_2$$
$$+ dNH_3$$

$a = 50; b = 100; c = 20; d = 5.$

$$C_{50}H_{100}O_{20}N_5 + 18.75H_2O \rightarrow 30.63CH_4 + 19.375CO_2 + 5NH_3$$

For methane gas:

$$(50 \times 12) + (100 \times 1) + (20 \times 16) + (5 \times 14)(waste) \rightarrow 30.63 \times (1 \times 12 + 4 \times 1)(CH_4)$$

$$1,090\,kg\ of\ waste \rightarrow 490\,kg\ CH_4$$

$$1t\ of\ waste \rightarrow (490 \times 1,000)/1,020 = 450\,kg\ CH_4$$

$$= 628\,m^3CH_4 (since\ density\ of\ CH_4 = 0.7167\,kg/m^3)$$

For carbon dioxide:

$$(50 \times 12) + (100 \times 1) + (20 \times 16) + (5 \times 14)(waste) \rightarrow 19.375 \times (1 \times 12 + 2 \times 16)(CO_2)$$

$$1,090\,kg\ of\ waste \rightarrow 853\,kg\ CO_2$$

$$1\ tonne\ of\ waste \rightarrow (853 \times 1,000)/1,020 = 783\,kg\ CO_2$$

$$= 400\,m^3CO_2 (since\ density\ of\ CO_2 = 1.9768\,kg/m^3)$$

Total biogas production per tonne of waste $= 628\,m^3 CH_4 + 400\,m^3 CO_2$

$= 1,028\,m^3$, biogas containing 61% CH_4 and 39% CO_2

Since only 80% is biodegraded during a 20 day period, total biogas potential

$$= 88\%\left(1,028\,m^3\right) = 822\,m^3/\text{tonne of waste}$$

Calorific value $= 23,000\,KJ/m^3 \times 822\,m^3 = 19 \times 10^6\,KJ/\text{tonne}$

Question B3: Assessment of Treatment Options for Municipal Solid Waste

a. Characterization

Table QB3 shows a typical composition of organic fraction of a sample of municipal solid waste. Determine the overall ash-free chemical composition of the waste.

b. Anaerobic digestion

Estimate the theoretical volume of methane gas and its calorific value that would be expected from the digestion of one tonne of waste described in Table B.1 above. Use density values for methane and carbon dioxide as 0.7167 and 1.9768 kg/m^3 respectively, and the calorific value of biogas as 23,000 KJ/m^3.

c. Aerobic composting

Determine the amount of air required to oxidize completely 1 tonne of the waste.

Assume that air contains 23.15% of oxygen by weight and has a density of 1.2928 kg/m^3.

TABLE B.1

Typical Composition of Organic Fraction of MSW

| Type | Mass (kg) | MC (%) | Ultimate Analysis (%Dry Mass) | | | | | |
			C	H	O	N	S	Ash
Food wastes	20	70	48.0	6.4	37.6	2.6	0.4	5.0
Garden wastes	9	60	47.8	6.0	38.0	3.4	0.3	4.5
Paper	40	6	43.5	6.0	44.0	0.3	0.2	6.0
Cardboard	6	5	44.0	5.9	44.6	0.3	0.2	5.0
Plastics	10	2	60.0	7.2	22.8	0.0	0.0	10.0
Wood	5	20	49.5	6.0	42.7	0.2	0.1	1.5
Others	10	8	26.3	3.0	2.0	0.5	0.2	68.0

Source: Tchobanoglous et al. 1993.

Solution B3

  a. Determination of the ash-free chemical composition of the wastes

  i. Compute the moisture content (MC) as illustrated in Table B.2

$$\text{Moisture content}(\%) = \left(\frac{x-y}{x}\right)100$$

where, x = initial mass of sample, and y = mass of sample after drying

$$MC = (100 - 75.9) = 24.1\,kg$$
$$= 24\%$$

  ii. Compute the chemical composition of the waste in mass as shown in Table B.3

TABLE B.2

Moisture Content

Type	Wet Mass (kg)	MC (%)	Dry mass (kg)
Food wastes	20	70	6
Garden wastes	9	60	3.6
Paper	40	6	37.6
Cardboard	6	5	5.7
Plastics	10	2	9.8
Wood	5	20	4.0
Other	10	8	9.2

TABLE B.3

Chemical Composition

Type	Mass (kg)	MC (%)	Dry Mass (kg)	Ultimate Analysis (kg)					
				C	H	O	N	S	Ash
Food wastes	20	70	6	2.88	0.38	2.26	0.16	0.02	0.30
Garden wastes	9	60	3.6	1.72	0.22	1.37	0.12	0.01	0.16
Paper	40	6	37.6	16.36	2.26	16.54	0.11	0.08	2.26
Cardboard	6	5	5.7	2.51	0.34	2.54	0.02	0.01	0.29
Plastics	10	2	9.8	5.88	0.71	2.23	0.00	0.00	0.98
Wood	5	20	4.0	1.98	0.24	1.71	0.01	0.004	0.06
Others	10	8	9.2	2.42	0.28	0.18	0.05	0.02	6.26
Moisture[a]			24.1	0.00	2.70	21.48	0.00	0.00	0.00
Total			100	33.75	7.13	48.26	0.47	0.14	10.31

[a] Convert the moisture content into hydrogen (H) and oxygen (O).

Conversion moisture it its hydrogen and oxygen constituents:

Molecular weight of water $(H_2O) = 18$

$$H = (2/18) \times 24.1 = 2.70 \, kg$$

$$O = (16/18) \times 24.1 = 21.48 \, kg$$

iii. Obtain elemental percentage mass composition as shown in Table B.4

iv. Determine molar composition of the elements as shown in Table B.5

v. Determine an approximate chemical formula with and without sulfur.

- Determine normalized mole ratios as shown in Table B.6
- Chemical formula with sulfur: $C_{702.5}H_{1765}O_{755}N_{7.5}S$ or $C_{703}H_{1765}O_{755}N_8S$
- Chemical formula without sulfur: $C_{93.5}H_{235}O_{100.7}N$ or $C_{945}H_{235}O_{101}N$

TABLE B.4

Elemental Mass Composition

Component	Mass (kg)	Percent (%) by Mass
Carbon (C)	33.75	33.75
Hydrogen (H)	7.13	7.13
Oxygen (O)	48.26	48.26
Nitrogen (N)	0.47	0.47
Sulfur (S)	0.14	0.14
Ash	10.31	10.31
Total	100	100

TABLE B.5

Molar Composition

Element	Mass (kg)	kg/mol	Moles
Carbon (C)	33.75	12.01	2.81
Hydrogen (H)	7.13	1.01	7.06
Oxygen (O)	48.26	16.00	3.02
Nitrogen (N)	0.47	14.01	0.03
Sulfur (S)	0.14	32.07	0.004
Ash	10.31	10.31	

TABLE B.6

Normalized Mole Ratios

Element	S = 1	N = 1
Carbon (C)	702.5	93.7
Hydrogen (H)	1,765.0	235.0
Oxygen (O)	755.0	100.7
Nitrogen (N)	7.5	1.0
Sulfur (S)	1.0	0

b. Anaerobic digestion

- Chemical composition of waste:

    The chemical formula with sulphur (i.e., $C_{702.5}H_{1765}O_{755}N_{7.5}S$), and using Equation 3.4 (Chapter 3). If sulfur is not considered (i.e., $C_{93.5}H_{235}O_{100.7}N$), Equation 3.3 will be used. *Both equations assume complete stabilization of the waste.*

    Using Equation 3.4:

$$C_aH_bO_cN_dS_e + 0.25(4a - b - 2c + 3d + 2e)H_2O \rightarrow 0.125(4a + b - 2c - 3d - 2e)CH_4$$

$$+ 0.125(4a - b + 2c + 3d + 2e)CO_2$$

$$+ dNH_3 + eH_2S$$

The coefficients are: $a = 702.5$; $b = 1,765$; $c = 755$; $d = 7.5$; $e = 1$. Using these coefficients, the resulting equation is:

$$C_{702.5}H_{1765}O_{755}N_{7.5}S \rightarrow 380CH_4 + 322CO_2 + 7.5NH_3 + H_2S$$

- Determination of the amount of mass of methane produced per tonne of waste:

$$(702.5 \times 12) + (1,765 \times 1) + (755 \times 16) + (7.5 \times 14) + (1 \times 32)(\text{waste}) \rightarrow$$

$$380 \times (1 \times 12 + 4 \times 1)(CH_4)$$

$$22,412 \text{ kg of waste} \rightarrow 6,080 \text{ kg CH}_4$$

$$1 \text{ tonne of waste} \rightarrow (6,080 / 22,412) \times 1,000 \text{ kg/tonne} = 271 \text{ kg CH}_4$$

Using a density value for methane of 0.7167 kg/m^3

$$\text{Volume of methane gas Mass/density} = \frac{271 \text{ kg/tonne}}{0.7167 \text{ kg/m}^3}$$

$$= 378 \text{ m}^3 CH_4/\text{tonne of waste}$$

- Determination of the mass of carbon dioxide produced per tonne of waste:

$$(702.5 \times 12) + (1,765 \times 1) + (755 \times 16) + (7.5 \times 14) + (1 \times 32)(\text{waste}) \rightarrow$$
$$322 \times (1 \times 12 + 2 \times 16)(CO_2)$$

$$22,412\,\text{kg of waste} \rightarrow 14,168\,\text{kg } CO_2$$

$$1\,\text{tonne of waste} \rightarrow (14,168 \,/\, 22,412) \times 1,000\,\text{kg/tonne} = 632\,\text{kg } CO_2$$

Using a density value for CO_2 of $1.9768\,\text{kg/m}^3$

$$\text{Volume of carbon dioxide Mass/density} = \frac{632\,\text{kg/tonne}}{1.9768\,\text{kg/m}^3}$$
$$= 320\,\text{m}^3 CO_2/\text{tonne of waste}$$

- The total amount of biogas production:

$$= 378\,\text{m}^3 CH_4 + 320\,\text{m}^3 CO_2, \text{ containing } 54\% \ CH_4 \text{ and } 46\% \ CO_2$$

- The calorific value waste:
  Using a calorific value for biogas of 23,000 KJ/m³

$$\text{The Theoretical calorific value} = 23,000\,\text{KJ/m}^3 \times 698\,\text{m}^3$$

$$= 16 \times 10^6\,\text{KJ/tonne of waste}$$

Actual biogas production:
The above calculations assume that all the organic components are readily biodegradable, which is not true in this case since the waste also contains plastics, wood, papers and cardboard, which can take relatively long time to breakdown. Hence, the realistic recoverable energy from the waste will be a fraction of the estimated theoretical value and this can be determined through BMP tests (see Chapter 4).

c. Aerobic composting
- Carbonaceous oxygen demand:
  Assume complete stabilization in the first instance. Oxygen demand for complete oxidation of the organic constituents can be obtained from Equation 1.10 (Chapter 1):

$$C_a H_b O_c N_d + 0.25(4a+b-2c-3d)O_2 \rightarrow aCO_2 + 0.5(b-3d)H_2O + dNH_3$$

Stoichiometric or chemical formula of waste without sulfur: $C_{93.5}H_{235}O_{100.7}N$

Using the coefficients: $a = 93.5$; $b = 235$; $c = 100.7$; $d = 1$; the resulting equation is:

$$C_{93.5}H_{235}O_{100.7}N + 101.5O_2 \rightarrow 93.5CO_2 + 117.5H_2O + NH_3$$

$$[(93.5 \times 12) + (235 \times 1) + (100.7 \times 16) + (1 \times 14)](\text{waste}) + (101.15 \times 16 \times 2)(O_2) \rightarrow$$

$$93.5 \times (1 \times 12 + 16 \times 2)(CO_2) + 117.5(1 \times 2 + 1 \times 16)(H_2O) + (1 \times 14 + 1 \times 3)(NH_3)$$

$$= 2982.2(\text{waste}) + 3236.8(O_2) \rightarrow 4,114(CO_2) + 3,185(H_2O) + 17(NH_3)$$

Oxygen required per ton of waste to stabilize organic matter = $(3,236.8/2,892.2) \times 1,000$ kg = $1,085$ kg

- Nitrogenous oxygen demand:

  Oxygen requirement for complete oxidation of ammonia can be obtained from Equation 1.11:

$$NH_3 + 2O_2 \rightarrow H_2O + HNO_3$$

Oxygen required per tonne to stabilize the ammonia:

$$NH_3 + 2O_2 \rightarrow H_2O + HNO_3$$

$$(1 \times 14 + 1 \times 3)(NH_3) + (2 \times 16 \times 2)(O_2) \rightarrow$$

$$(1 \times 2 + 1 \times 16)(H_2O) + (1 \times 1 + 14 \times 1 + 3 \times 16)(HNO_3)$$

$$17(NH_3) + 64(2O_2) \rightarrow 18(H_2O) + 63(HNO_3)$$

Oxygen required per tonne of waste to stabilize ammonia = $(64/2,982.2) \times 1,000$ kg = 22 kg

- The total mass of oxygen required:

$$O_{2,\text{total}} = 1,085 + 22 = 1,107 \text{ kg/tonne of waste}$$

- The mass of air required:

  Assume air contains 23.15% oxygen by weight and density of air is $1.2928\,\text{kg/m}^3$.

$$\text{Air}_{\text{mass}} = \frac{1,107\,\text{kg/tonne}}{23.15\%} = 4,782\,\text{kg/tonne}$$

- The volume of air required:

$$Air_{volume} = Mass/density = (4,782\,kg/tonne)/(1.2928\,kg/m^3)$$

$$= 3,699\,m^3/tonne\ of\ waste$$

Note:

i. The above calculations assume that all the organic materials are readily biodegradable, however, some (e.g. paper cardboard and wood) will take relatively long time to breakdown. Hence, the actual amount of air required for complete breakdown of the waste will be a fraction of the estimated theoretical value.

ii. The theoretical volume of air also assumes 100% oxygen transfer efficiency. Oxygen transfer efficiency depends on many factors, including temperature, air pumping system and waste characteristics. Generally, less than 10% efficiency is achievable. Taking this into consideration:

$$Actual\ amount\ of\ air\ required = (Air_{volume}, m^3/tonne\ of\ waste)/$$
$$(O_2 transfer\ efficiency,\ \%)$$

The actual amount is used in choosing appropriate air supply facilities.

References and Further Reading

Abbasi, T., S. M. Tauseef and S. A. Abbasi. 2012. Anaerobic digestion for global warming control and energy generation: an overview. *Renewable and Sustainable Energy Reviews* 16, 5, 3228–3242.

Abdullahi, Y. A. 2008. *Optimization of anaerobic digestion of organic solid waste for the production of quality compost for soil amendment.* PhD Thesis. University of Abertay, Dundee.

Abdullahi, Y. A., J. C. Akunna, N. A. White, P. D. Hallett and R. Wheatley. 2008. Investigating the effects of anaerobic and aerobic post-treatment on quality and stability of organic fraction of municipal solid waste as soil amendment. *Bioresource Technology* 99, 8631–8636.

Akunna, J. C. and G. M. Walker. 2017. Co-Products from malt whisky production and their utilisation. In *The Alcohol Textbook*, 6th ed. eds. G. M. Walker, C. Abbas, W. M. Ingledew and E. Pilgrim, 529–537. Duluth: Ethanol Technology Institute.

Akunna, J. C. 1995. Denitrification in anaerobic digesters: a review of recent studies. In *Proceedings, 50th Annual Purdue Industrial Waste Conference*, 395–403. Chelsea: Ann Arbor Press, Inc.

Akunna, J. C. 2015. Anaerobic treatment of brewery wastes. In *Brewing Microbiology: Managing Microbes, Ensuring Quality and Valorising Waste*, ed. A. E. Hill, 407–424. Cambridge: Woodhead Publishing, Elsevier.

Akunna, J. C. and A. Hierholtzer. 2016. Co-digestion of terrestrial plant biomass with marine macro-algae for biogas production. *Biomass and Bioenergy* 93, 137–143.

Akunna, J. C. and B. Kwiatkowska. 2012. Investigating factors affecting anaerobic digestion of *Ascophylum nodosum* species of seaweed. *Platform Presentation & Proceedings, Fourth International Symposium on Energy from Biomass and Waste*, November 12–15, Venice.

Akunna, J. C. and J. Bartie. 2014. Wastewater treatment, infrastructure and design. In *Water Resources in the Built Environment—Management Issues and Solutions*, eds. C. Booth and S. Charlesworth, 350–370. Chichester: John Wiley & Sons Ltd.

Akunna, J. C. and M. Clark. 2000. Performance of a granular-bed anaerobic baffled reactor (GRABBR) treating whisky distillery wastewater. *Bioresource Technology* 73, 3, 257–261.

Akunna, J. C., C. Bizeau and R. Moletta. 1992. Denitrification in anaerobic digesters: possibilities and influence of wastewater COD/N-NOx ratio. *Environmental Technology* 13, 825–836.

Akunna, J. C., C. Bizeau and R. Moletta. 1993. Nitrate and nitrite reductions with anaerobic sludge using various carbon sources: glucose, glycerol, acetic acid, lactic acid and methanol. *Water Research* 27, 1303–1312.

Akunna, J. C., C. Bizeau and R. Moletta. 1994. Nitrate reduction by anaerobic sludge using glucose at various nitrate concentrations: ammonification, denitrification and methanogenic activities. *Environmental Technology* 15, 41–49.

Akunna, J. C., E. Kuzmanova and O. Mašek. 2015. Anaerobic digestion of slow pyrolysis liquids from various types of biomass. In *Proceedings, 14th World Congress on Anaerobic Digestion*, November 15–18, Viña Del Mar: International Water Association.

Akunna, J. C., K. Hasan and K. Kerr. 2009. Estimating methane production potential of an old municipal landfill. *The Journal of Solid Waste Technology and Management* 35, 3, 156–161.

Akunna, J. C., N. Bernet and R. Moletta. 1998. Effect of nitrate on methanogenesis at low redox potential. *Environmental Technology* 19, 1249–1254.

Akunna, J. C., Y. A. Abdullahi and N. A. Stewart. 2007. Anaerobic digestion of municipal solid wastes containing variable proportions of waste types. *Water Science & Technology* 56, 8, 143–149.

Akunna, J., C. Bizeau, R. Moletta, N. Bernet and A. Héduit. 1994. Combined organic carbon and complete nitrogen removal using anaerobic and aerobic upflow filters. *Water Science & Technology* 30, 12, 297–306.

Al Qarni, H. 2015. *Investigating human pharmaceutical compounds present in municipal and hospital wastewater and options for their removal.* PhD Thesis. University of Abertay, Dundee.

Al Qarni, H., P. Collier, J. O'Keeffe and J. Akunna. 2016. Investigating the removal of some pharmaceutical compounds in hospital wastewater treatment plants operating in Saudi Arabia. *Environmental Science and Pollution Research* 23, 13, 13003–13014.

Al-Alm Rashed, I. G., J. Akunna, M. M. El-Halwany and A. F. F. Abou Atiaa. 2010. Improvement in the efficiency of hydrolysis of anaerobic digestion in sewage sludge by the use of enzymes. *Desalination and Water Treatment* 21, 280–285.

Angelidaki, I. and W. Sanders. 2004. Assessment of the anaerobic biodegradability of macropollutants. *Reviews in Environmental Science and Bio/Technology* 3, 117–129.

Angelidaki, I., M. Alves, D. Bolzonella, L. Borzacconi, J. L. Campos, A. J. Guwy, S. Kalyuzhnyi, P. Jenicek and J. B. van Lier. 2009. Defining the biomethane potential (BMP) of solid organic wastes and energy crops: a proposed protocol for batch assays. *Water Science and Technology* 59, 5, 927–934.

Appels, L., J. Baeyens, J. Degrève and R. Dewil. 2008. Principles and potential of the anaerobic digestion of waste-activated sludge. *Progress in Energy and Combustion Science* 34, 755–781.

Appels, L., S. Houtmeyers, J. Degrève, J. Van Impe and R. Dewil. 2013. Influence of microwave pre-treatment on sludge solubilisation and pilot scale semi-continuous anaerobic digestion. *Bioresource Technology* 128, 598–603.

Ariunbaatar, J., A. Panico, G. Esposito, F. Pirozzi and P. N. Lens. 2014a. Pre-treatment methods to enhance anaerobic digestion of organic solid waste. *Applied Energy* 123, 143–156.

Ariunbaatar, J., A. Panico, L. Frunzo, G. Esposito, P. N. Lens and F. Pirozzi. 2014b. Enhanced anaerobic digestion of food waste by thermal and ozonation pre-treatment methods. *Journal of Environmental Management* 146, 142–149.

Attal, A., J. Akunna, P. Camacho, P. Salmon and I. Paris. 1992. Anaerobic degradation of municipal wastes in landfill. *Water Science & Technology* 25, 7, 243–253.

Baloch, M. I. 2004. *Carbon and nitrogen removal in a granular bed baffled reactor.* PhD Thesis. University of Abertay, Dundee.

Baloch, M. I. and J. C. Akunna. 2003a. Granular bed baffled reactor (GRABBR): a solution to a two-phase anaerobic digestion system. *American Society of Civil Engineers (ASCE) Journal of Environmental Engineering* 29, 11, 1015–1021.

Baloch, M. I. and J. C. Akunna. 2003b. Effect of rapid hydraulic shock loads on the performance of granular bed baffled reactor. *Environmental Technology* 24, 361–368.

Baloch, M. I., J. C. Akunna and P. J. Collier. 2004. Effectiveness of a plug flow granular bed multi-stage anaerobic system for carbon and nitrate removal. In *Proceedings, 10th World Congress of Anaerobic Digestion* 2004 on Anaerobic Bioconversion for Sustainability, August 29–September 2, 354–360, Montreal.

Baloch, M. I., J. C. Akunna and P. J. Collier. 2006a. Carbon and nitrogen removal in a granular bed baffled reactor. *Environmental Technology* 27, 2, 201–208.

Baloch, M. I., J. C. Akunna and P. J. Collier. 2006b. Assessment of morphology for anaerobic-granular particles. *Water Environment Research* 78, 6, 643–646.

Baloch, M. I., J. C. Akunna and P. J. Collier. 2007. The performance of a phase separated granular bed bioreactor treating brewery wastewater. *Bioresource Technology* 98, 1849–1855.

Baloch, M. I., J. C. Akunna, M. Kierans and P. J. Collier. 2008. Structural analysis of anaerobic granules in a phase separated reactor by electron microscopy. *Bioresource Technology* 99, 922–929.

Banks, C. J., Y. Zhang, Y. Jiang and S. Haven. 2012. Trace element requirements for stable food waste digestion at elevated ammonia concentrations. *Bioresource Technology* 104, 127–135.

Barber, 2005. Anaerobic digester foaming: causes and solutions. *Water* 21, 45–49 (February).

Barber, W. P. and D. C. Stuckey. 1999. The use of anaerobic baffled reactor for wastewater treatment: a review. *Water Research* 33, 11, 1559–1578.

Bashir, B. H. and A. Martin. 2004a. Effect of calcium and potassium on sodium inhibition to methanogens in anaerobic treatment processes. *Electronic Journal of Environmental, Agricultural and Food Chemistry* 3, 769–776.

Bashir, B. H. and A. Martin. 2004b. Effect of calcium and magnesium on sodium inhibition to methanogens in anaerobic treatment processes. *Electronic Journal of Environmental, Agricultural and Food Chemistry* 4, 827–834.

Bennett, J., G. Walker, D. Murray, J. Akunna and A. Wardlaw. 2015. Avenues for bioenergy production using malt distillery co-products. In *Proceedings, Worldwide Distilled Spirits Conference*, September 8–11, 303–312, Glasgow: The Institute of Distilling & Brewing.

Bernet, N., N. Delgenes, J. C. Akunna, J.-P. Delgenes and R. Moletta. 2000. Combined anaerobic-aerobic SBR for the treatment of piggery wastewater. *Water Research* 34, 2, 611–619.

Bremner, D. 1990. *Historical Introduction to Sonochemistry*. London: JAI Press Ltd.

Brière, F. G., P. Béron and R. Haulser. 1994. Influence of suspended solids in ozonation process. *IOA Regional Conference*, S7.43–S7.52, Zurich.

Bruton, T., H. Lyons, Y. Lerat, M. Stanley and B. M. Rasmussen. 2009. A review of the potential of marine algae as a source of biofuel in Ireland. Report prepared for Sustainable Energy Ireland by BioXL, Shannon Applied Biotechnology Centre, European Research Centre for Algae (CEVA), Scottish Association for Marine Science (SAMS) and National Environmental Research Institute (NERI).

Cammarota, M. C., G. A. Teixeira and D. M. G. Freire. 2001. Enzymatic pre-hydrolysis and anaerobic degradation of wastewater with high oil contents. *Biotechnology Letters* 23, 1591–1595.

Chen, Y., J. Cheng and K. S. Creamer. 2008. Inhibition of anaerobic digestion process: a review. *Bioresource Technology* 99, 4044–4064.

Couth, R. and C. Trois. 2012. Cost-effective waste management through composting in Africa. *Waste Management* 32, 12, 2518–2525.

Crites, R. and G. Tchobanoglous. 1998. *Small and Decentralized Wastewater Management Systems*, New York: McGraw-Hill.

Delgenès, J. P., V. Penaud and R. Moletta. 2003. Pre-treatments for the enhancement of anaerobic digestion of solid wastes. In *Biomethanization of Organic Fraction of Municipal Solid Waste*, ed. J. Mata-Alvarez, 201–228. Cornwall: IWA Publishing.

Demirel, B. and P. Scherer. 2011. Trace element requirements of agricultural biogas digesters during biological conversion of renewable biomass to methane: review. *Biomass Bioenergy* 35, 992–998.

Deublein, D. and A. Steinhauser. 2008. *Biogas from Waste and Renewable Resources*. Weinheim: Willey-VCH Verlag GmbH & Co. KGaA.

Dhouib, A., M. Ellouz, F. Aloui and S. Sayadi. 2006. Effect of bio-augmentation of activated sludge with white-rot fungi on olive mill wastewater detoxification. *Letters in Applied Microbiology* 42, 405–411.

Diaz, A. B., M. M. de Souza Moretti, C. Bezerra-Bussoli, C. da Costa Carreira Nunes, A. Blandino, R. da Silva and E. Gomes. 2015. Evaluation of microwave-assisted pre-treatment of lignocellulosic biomass immersed in alkaline glycerol for fermentable sugars production. *Bioresource Technology* 185, 316–323.

Eckenfelder, W. W. Jr. 1989. *Industrial Water Pollution Control*. Singapore: McGraw-Hill.

Eder, B. and H. Schulz. 2006. *Biogas-Praxis, Grundlagen, Planung. Anlagenbau, Beispiele Wirtschaftlichkeit*. 3rd ed. Staufen bei Freiburg: ökobuch Verlag.

Environmental Protection Agency (EPA). 1979. Water Related Environmental Fate of 129 Priority Pollutants. EPA-440/4-79-029, Washington, DC.

Facchin, V., C. Cavinato, F. Fatone, P. Pavan, F. Cecchi and D. Bolzonella. 2013. Effect of trace element supplementation on the mesophilic anaerobic digestion of food waste in batch trials: the influence of inoculum origin. *Biochemical Engineering Journal* 70, 71–77.

Feijoo, G., M. Soto, R. Méndez and J. M. Lema. 1995. Sodium inhibition in the anaerobic digestion process: antagonism and adaptation phenomena. *Enzyme and Microbial Technology* 17, 180–188.

Friedrich, E. and C. Trois. 2011. Quantification of greenhouse gas emissions from waste management processes for municipalities—a comparative review focusing on Africa. *Waste Management* 31, 7, 1585–1596.

Geib, S. M., T. R. Filley, P. G. Hatcher, K. Hoover, J. E. Carlson, M. Jimenez-Gasco Mdel, A. Nakagawa-Izumi, R. L. Sleighter and M. Tien. 2008. Lignin degradation in wood-feeding insects. *Proceedings of the National Academy of Sciences USA* 105, 35, 12932–12937.

German Solar Energy Society (DGS) and Ecofys. 2005. *Planning and Installing Bioenergy Systems: A Guide for Installers, Architects and Engineers*. 53–100. London: Earthscan.

Gevaert, F., M.–A. Janquin and D. Davoult. 2008. Biometrics in *Laminaria digitata*: a useful tool to assess biomass, carbon and nitrogen contents. *Journal of Sea Research* 60, 215–219.

Ghosh, A. and B. C. Bhayyacharyya. 1999. Biomethanation of white rotted and brown rotted rice straw. *Bioprocess Engineering* 20, 4, 297–302.

Gilbert, E. 1983. Investigations on the changes of biological biodegradability of single substances induced y ozonation. *Ozone: Science and Engineering* 5, 137–149.

Gray, N. F. 2010. *Water Technology: An Introduction for Environmental Scientists and Engineers*. Amsterdam: Elsevier.

Guiot, S. R., R. Cimpoia and G. Carayon. 2011. Potential of wastewater-treating anaerobic granules for biomethanation of synthesis gas. *Environmental Science & Technology* 44, 5, 2006–2012.

Guiot, S. R., R. Cimpoia, S. Sancho Navarro, A. Prudhomme and M. Filiatrault. 2013. Anaerobic digestion for bio-upgrading syngas into renewable natural gas (methane). In *Proceedings, 13th World Congress on Anaerobic Digestion*, June 25–28, Santiago de Compostela: International Water Association.

Gujer, W. and A. J. B. Zehnder. 1983. Conversion processes in anaerobic digestion. *Water Science and Technology* 15, 127–167.

Hall, E. R. 1992. Anaerobic treatment of wastewater in suspended growth and fixed film processes. In *Design of Anaerobic Processes for the Treatment of Industrial and Municipal Wastes*, eds. J. F. Malina and F. G. Pohland, 41–118. Lancaster: Technomic Publishing Company Inc.

Hansen, T. L., J. E. Schmidt, I. Angelidaki, E. Marca, J. I. Jansen, H. Mosbaek and T. H. Christensen. 2004. Method for determination of methane potentials of solid organic waste. *Waste Management* 24, 4, 393–400.

Hartmann, H. and B. K. Ahring. 2006. Strategies for the anaerobic digestion of the organic fraction of municipal solid waste: an overview. *Water Science and Technology* 53, 8, 7–22.

Hartmann, H., I. Angelidaki and B. K. Ahring. 2003. Co-digestion of the organic fraction of municipal waste with other waste types. In *Biomethanization of Organic Fraction of Municipal Solid Waste*, ed. J. Mata-Alvarez, 181–197. Cornwall: IWA Publishing.

Hendriks, A. T. W. M. and G. Zeeman. 2009. Pretreatments to enhance the digestibility of lignocellulose biomass. *Bioresource Technology* 100, 10–18.

Hierholtzer, A. 2013. *Investigating factors affecting the anaerobic digestion of seaweed: modelling and experimental approaches*. PhD Thesis. University of Abertay, Dundee.

Hierholtzer, A. and J. C. Akunna. 2012. Modelling sodium inhibition on the anaerobic digestion process. *Water Science & Technology* 66, 7, 1656–1573.

Hierholtzer, A. and J. C. Akunna. 2014. Modelling start-up performance of anaerobic digestion of saline-rich macro-algae. *Water Science & Technology* 69, 10, 2059–2064.

Hierholtzer, A., L. Chatellard, M. Kierans, J. C. Akunna and P. J. Collier. 2013. The impact and mode of action of phenolic compounds extracted from brown seaweed on mixed anaerobic microbial cultures. *Journal of Applied Microbiology* 114, 4, 964–973.

Hills, D. J. and K. Nikano. 1984. Effects of particle size on anaerobic digestion of tomato solid wastes. *Agricultural Wastes* 10, 285–295.

Hjorth, M., K. Gränitz, A. P. S. Adamsen and H. B. Møller. 2011. Extrusion as a pretreatment to increase biogas production. *Bioresource Technology* 102, 4989–4994.

Hodgson, J., C. Laugero, R. Leduc, M. Asther and S. R. Guiot. 1998. Fungal pretreatment by *Phanerochaete chrysosporium* to reduce the inhibition of methanogenesis by dehydroabietic acid. *Applied Microbiology and Biotechnology* 49, 538–544.

Houtmeyers, S., J. Degrève, K. Willems, R. Dewil and L. Appels. 2014. Comparing the influence of low power ultrasonic and microwave pre-treatments on the solubilisation and semi-continuous anaerobic digestion of waste activated sludge. *Bioresource Technology* 171, 44–49.

Hübner, T. and J. Mumme. 2015. Integration of pyrolysis and anaerobic digestion—use of aqueous liquor from digestate pyrolysis for biogas production. *Bioresource Technology* 183, 86–92.

International Energy Agency (IEA) Bioenergy. 2006. Biogas upgrading to vehicle fuel standards and grid injection, eds. M. Persson, O. Jönsson and A. Wellinger. *IEA Bioenergy Task 37—Energy from Biogas*, IEA Bioenergy.

International Energy Agency (IEA) Bioenergy. 2009. Biogas upgrading technologies—developments and innovations, eds. A. Petersson and A. Wellinger. *IEA Bioenergy Task 37—Energy from Biogas*, IEA Bioenergy.

International Energy Agency (IEA) Bioenergy. 2010. Utilisation of digestate from biogas plants as biofertiliser, eds. C. T. Lukehurst, P. Frost and T Al Seadi. *IEA Bioenergy Task 37—Energy from Biogas*, IEA Bioenergy.

International Energy Agency (IEA) Bioenergy. 2011. Biogas from energy crop digestion, eds. J. Murphy, R. Braun, P. Weiland and A. Wellinger. *IEA Bioenergy Task 37—Energy from Biogas and Landfill Gas*, IEA Bioenergy.

International Energy Agency (IEA) Bioenergy. 2013. Process monitoring in biogas plants, ed. B. Drosg. *IEA Bioenergy Task 37—Energy from Biogas and Landfill Gas*, IEA Bioenergy.

International Energy Agency (IEA) Bioenergy. 2014a. A perspective on the potential role of biogas in smart energy grids, eds. T. Persson, J. Murphy, A-K Jannasch, E. Ahern, J. Liebetrau, M. Trommler and J. Toyama. *IEA Bioenergy Task 37—Energy from Biogas*, IEA Bioenergy.

International Energy Agency (IEA) Bioenergy. 2014b. Pre-treatment of feedstock for enhanced biogas production, eds. L. F. R. Montgomery and G. Bochmann. *IEA Bioenergy Task 37—Energy from Biogas*, IEA Bioenergy.

Jayachandra, T., C. Venugopal and K. A. Anu Appaiah. 2011. Utilization of phytotoxic agro waste: coffee cherry husk through pre-treatment by the ascomycetes fungi *Mycotypha* for biomethanation. *Energy for Sustainable Development* 15, 104–108.

Jung, F., M. C. Cammarota and D. M. G. Freire. 2002. Impact of enzymatic pre-hydrolysis on batch activated sludge systems dealing with oily wastewaters. *Biotechnology Letter* 24, 1797–1802.

Kasper, H. F. and K. Wuhrann. 1978. Kinetic parameters and relative turn-overs of some important catabolic reactions in digesting sludge. *Applied and Environmental Microbiology* 36, 1–7.

Kiely, G. 1997. *Introduction to Environmental Engineering*. New York: McGraw-Hill.

Knappert, D., H. Grethlein and A. Converse. 1981. Partial acid hydrolysis of popular wood as a pre-treatment for enzymatic hydrolysis. *Biotechnology and Bioengineering Symposium* 11, 67–77.

Kong, F., C. Engler and E. Soltes. 1992. Effects of cell-wall acetate, xylan backbone and lignin on enzymatic hydrolysis of aspen wood. *Applied Biochemistry and Biotechnology* 34, 23–35.

Kuzmanova, E., N. Zhelev and J. C. Akunna. 2018. Effect of liquid nitrogen pre-treatment on various types of wool fibres for biogas production. *Heliyon* 4, 5, e00619.

Kwiatkowska, B., J. Bennett, J. Akunna, G. M. Walker and D. H. Bremner. 2011. Stimulation of bioprocesses by ultrasound. Review article. *Biotechnology Advances* 29, 6, 768–780.

Laureano-Perez, L., F. Teymouri, H. Alizadeh and B. E. Dale. 2005. Understanding factors that limit enzymatic hydrolysis of biomass. *Applied Biochemistry and Biotechnology* 121–124, 1081–1099.

Lossie, U. and P. Pütz. 2008. Practice report: Laboratory analysis, titration. Targeted control of biogas plants with the help of FOS/TAC. Hach Lange. Retrieved on December 12, 2017, from https://tr.hach.com/asset-get.download.jsa?id=25593611361.

Malina, J. F. 1992. Anaerobic sludge digestion. In *Design of Anaerobic Processes for the Treatment of Industrial and Municipal Wastes*, eds. J. F. Malina and F. G. Pohland, 167–212. Lancaster: Technomic Publishing Company Inc.

Malina, J. F. and F. G. Pohland. 1992. *Design of Anaerobic Processes for the Treatment of Industrial and Municipal Wastes.* Lancaster: Technomic Publishing Company Inc.

Mallick, P., J. C. Akunna and G. M. Walker. 2010a. Benefits of enzymatic pre-treatment of intact yeast cells for anaerobic digestion of distillery pot ale. In *Distilled Spirits— Proceedings of the Worldwide Conference on Distilled Spirits*, eds. G. Walker and P. Hughes, 197–201. Nottingham: Nottingham University Press.

Mallick P., J. C. Akunna and G. Walker. 2010b. Anaerobic digestion of distillery spent wash: influence of enzymatic pre-treatment of intact yeast cells. *Bioresource Technology*, 101, 1681–1685.

Massé, L., D. I. Massé and K. J. Kennedy. 2003. Effect of hydrolysis pre-treatment on fat degradation during anaerobic digestion of slaughterhouse wastewater. *Process Biochemical* 38, 1365–1372.

Massé, L., K. J. Kennedy and S. Chou. 2001. Testing of alkaline and enzymatic hydrolysis pre-treatment for oil particles in slaughterhouse wastewater. *Bioresource Technology* 77, 145–155.

Mata-Alvarez, J. 2003. Fundamentals of the anaerobic digestion process. In *Biomethanization of Organic Fraction of Municipal Solid Waste*, ed. J. Mata-Alvarez, 1–20. Cornwall: IWA Publishing.

McCarty, P. L. 1964. Anaerobic waste treatment fundamentals, Parts, 1, 2, 3 and 4. *Public Works*, 95; 9, 107–112; 10, 123–126; 11, 91–94; 12, 95–99.

McCarty, P. L. 1968. Anaerobic treatment of soluble wastes. In *Advances in Water Quality Improvement*, ed. W. W. Eckenfelder, Jr. Austin: University of Texas Press.

McDonnell, G. E. 2007. Chemical disinfection. In *Antisepsis, Disinfection, and Sterilization: Types, Action, and Resistance*, ed. G. E. McDonnell, 96–148. Washington, DC: ASM Press.

McDonnell, G. E. and A. D. Russell. 1999. Antiseptics and disinfectants: activity, action, and resistance. *Clinical Microbiology Reviews* 12, 1, 147–179.

Metcalf & Eddy. 2014. *Wastewater Engineering: Treatment, Disposal and Reuse.* 5th ed. New York: McGraw-Hill.

Montgomery, L. F. R. and G. Bochmann. 2014. Pre-treatment of feedstocks for enhance biogas production, Energy Technology Network. International Energy Agency (IEA) Bioenergy.

Mucyo, S. 2013. *Analysis of key requirements for effective implementation of biogas technology for municipal solid waste management in sub-Saharan Africa. A case study of Kigali City.* Rwanda. PhD Thesis. University of Abertay, Dundee.

Mucyo, S. and J. C. Akunna. 2014. Municipal waste collection for effective recovery and reuse of organic fractions. Case study: Kigali, Rwanda. Proceedings, II International Africa Sustainable Waste Management Conference, International Solid Waste Association (ISWA), Association of Portuguese Sanitary and Environmental Engineering (APESB) and the Association of Ecologists and Environmentalists of Angola (AEEA). Luanda, Angola, April 22–24.

Mustafa, A. M., T. G. Poulsen, Y. Xia and K. Sheng. 2017. Combination of fungal and milling pre-treatments for enhancing rice straw biogas production during steady state anaerobic digestion. *Bioresource Technology* 224, 174–182.

Narkis, N. and M. Schneider-Rotel. 1980. Evaluation of ozone induced biodegradability of wastewater treatment plant effluent. *Water Research* 14, 929–939.

Neis, U., K. Nickel and A. Tiehm. 2001. Ultrasonic disintegration of sewage sludge for enhanced anaerobic biodegradation. In *Advances in Sonochemistry*, eds. T. J. Mason and A. Tiehm, 6, 59–90.

Nickel, K. and U. Neis. 2007. Ultrasonic disintegration of biosolids for improved biodegradation. *Ultrasonics Sonochemistry* 14, 450–455.

Nielsen, H. B., Z. Mladenovska and B. K. Ahring. 2007. Bio-augmentation of a two-stage thermophilic (68°C/55°C) anaerobic digestion concept for improvement of the methane yield from cattle manure. *Biotechnology and Bioengineering* 96, 6, 1638–1643.

Nizami, A.-S., A. Orozco, E. Groom, B. Dieterich and J. D. Murphy. 2012. How much gas we get from grass? *Applied Energy* 92, 783–790.

Obata, O. 2015. *Evaluating microbial population dynamics and system performance during anaerobic digestion of marine materials*. PhD Thesis. University of Aberdeen.

Obata, O., J. C. Akunna and G. Walker. 2015. Hydrolytic effects of acid and enzymatic pre-treatment on the anaerobic biodegradability of *Ascophyllum nodosum* and *Laminaria digitata* species of brown seaweed. *Biomass and Bioenergy*, 80, 140–146.

Ollivier, B., P. Caumette, J. L. Garcia and R. A. Mah. 1994. Anaerobic bacteria from hypersaline environments. *Microbiological Reviews* 58, 27–38.

Onyeche, T. I., O. Schläfer, H. Bormann, C. Schröder and M. Sievers. 2002. Ultrasonic cell disruption of stabilised sludge with subsequent anaerobic digestion. *Ultrasonics* 40, 31–35.

Onyia, C. O., A. M. Uyu, J. C. Akunna, N. A. Norulani and A. K. Omar. 2001. Increasing the fertilizer value of palm oil mill sludge: bioaugmentation in nitrification. *Water Science & Technology* 44, 10, 157–162.

Owen, W. F., D. C. Stuckey, J. B. Healy Jr, L. Y. Young and P. L. McCarty. 1979. Bioassays for monitoring biochemical methane potential and anaerobic toxicity. *Water Research* 13, 6, 485–492.

Ozbayram, E. G., S. Kleinsteuber, S. M. Nikolausz, B. Ince and O. Ince. 2016. Effect of bio-augmentation by cellulolytic bacteria enriched from sheep rumen on methane production from wheat straw. *Anaerobe* 46, 122–130.

Parkin, G. F. and W. F. Owen. 1986. Fundamental of anaerobic digestion of wastewater sludges. *Journal of Environmental Engineering* 112, 5, 887–920.

Peavy, H. S., Rowe, D. R. and Tchobanoglous, G. 1985. *Environmental Engineering*. 285–291. New York: McGraw Hill.

Peng, X., R. A. Börner, I. A. Nges and J. Liu. 2014. Impact of bioaugmentation on biochemical methane potential for wheat straw with addition of *Clostridium cellulolyticum*. *Bioresource Technology* 152, 567–571.

Pérez-Elvira, S. I, L. C. Ferreira, A. Donoso-Bravo, M. Fdz-Polanco and F. Fdz-Polanco. 2010. Full-stream and part-stream ultrasound treatment effect on sludge anaerobic digestion. *Water Science Technology* 61, 1363–1372.

Pilli, S., P. Bhunia, S. Yan, R. J. Leblanc, R. D. Tyagi and R. Y. Surampalli. 2011. Ultrasonic pretreatment of sludge: a review. *Ultrasonic Sonochemistry* 18, 1–18.

Pohland, F. G. 1992. Anaerobic treatment: fundamental concepts, applications, and new horizons. In *Design of Anaerobic Processes for the Treatment of Industrial and Municipal Wastes*, eds. J. F. Malina and F. G. Pohland, 1–40. Lancaster: Technomic Publishing Company Inc.

Polprasert, C. 2007. *Organic Waste Recycling: Technology and Management.* 145–206. London: IWA Publishing.

Raposo, F., M. A. de la Rubia, V. Fernández-Cegrí and R. Borja. 2012. Anaerobic digestion of solid organic substrates in batch mode: an overview relating to methane yields and experimental procedures. *Renewable and Sustainable Energy Reviews* 16, 1, 861–877.

Rich, L. G. 1963. *Unit Processes of Environmental Engineering.* New York: John Wiley.

Risberg, K., L. Sun, L. Levén, S. J. Horn and A. Schnürer. 2013. Biogas production from wheat straw and manure—impact of pre-treatment and process operating parameters. *Bioresource Technology* 149, 232–237.

Rosato, M. A. 2018. *Managing Biogas Plants: A Practical Guide.* Boca Raton, FL: CRC Press, Taylor & Francis Group.

Ross, A. B., J. M. Jones, M. L. Kubacki and T. Bridgeman. 2008. Classification of macroalgae as fuel and its thermochemical behaviour. *Bioresource Technology* 99, 6494–6504.

Saha, B. C. 2003. Hemicellulose bioconversion. *Journal of Industrial Microbiology and Biotechnology* 30, 5, 279–291.

Sawyer, C. N., P. L. McCarty and G. F. Parkin. 1994. *Chemistry for Environmental Engineering.* New York: McGraw-Hill.

Schell, D. and C. Hardwood. 1994. Milling of lignocellulostic biomass. *Applied Biochemistry and Biotechnology* 45, 159–168.

Shanmugam, A. and J. C. Akunna. 2008. Comparison of the performance of GRABBR and UASB for the treatment of low strength wastewaters. *Water Science & Technology* 58, 1, 225–232.

Shanmugam, A. and J. C. Akunna. 2010. Modelling head losses in granular bed anaerobic baffled reactors at high flows during start-up. *Water Research* 44, 5474–5480.

Sharma, S. K., L. M. Mishra, M. P. Sharma and J. S. Saini. 1988. Effect of particle size on biogas generation from biomass residues. *Biomass* 17, 251–263.

Show, K., T. Mao and D. J. Lee. 2007. Optimisation of sludge disruption by sonication. *Water Research* 41, 4741–4747.

Song, J. J., N. Takeda and M. Hiraoka. 1992. Anaerobic treatment of sewage treated by catalytic wet oxidation process in upflow anaerobic blanket reactors. *Water Science & Technology* 26, 3–4, 867–875.

Sosnowski, P., A. Wieczorek and S. Ledakiwicz. 2003. Anaerobic co-digestion of sewage sludge and organic fraction of municipal solid wastes. *Advances in Environmental Research* 7, 609–616.

Speece, R. E. 2008. *Anaerobic Biotechnology and Odor/Corrosion Control.* 246–249. Nashville, TN: Archae Press.

Stamatelatou, K., G. Antonopoulou and P. Michailides. 2014. Biomethane and bihydrogen production via anaerobic digestion/fermentation. In *Advances in Biorefineries*, ed. K. Waldron, 478–524. Cambridge: Elsevier.

Stenile, E. 1993. Sludge treatment and disposal systems for rural areas in Germany. *Water Science and Technology* 27, 9, 159–171.

Stronach, S. M., T. Rudd and J. N. Lester. 1986. *Anaerobic Digestion Processes in Industrial Wastewater Treatment*. Berlin: Springer-Verlag.

Sun, Y. and J. Cheng. 2002. Hydrolysis of lignocellulosic materials for ethanol production. A review. *Bioresource Technology* 83, 1–11.

Taherzadeh, M. J. and K. Karimi. 2008. Pretreatment of lignocellulosic wastes to improve ethanol and biogas production: a review. *International Journal of Molecular Sciences* 9, 1621–1651.

Tchobanoglous, G. and E. Schroeder. 1987. *Water Quality*. Reading, MA: Addison Wesley.

Tchobanoglous, G., H. Thiesen and S. Vigil. 1993. *Integrated Solid Waste Management*. 611–670. New York: McGraw-Hill.

Tiehm, A., K. Nickel and U. Neis. 1997. The use of ultrasound to accelerate the anaerobic digestion of sewage sludge. *Water Science and Technology* 36, 11, 121–128.

Tiehm, A., K. Nickel, M. Zellhorn and U. Neis. 2001. Ultrasonic waste activated sludge disintegration for improving anaerobic stabilization. *Water Research* 35, 2003–2009.

Tota-Maharaj, K., J. Akunna and D. Cheddie. 2017. Municipal wastewater treatment and associated bioenergy generation using anaerobic granular bed baffled reactor. *Journal of The Association of Professional Engineers of Trinidad and Tobago* 45, 1, 11–19.

Troschinetz, A. and A. M. Mihelcic. 2009. Sustainable recycling of municipal solid waste in developing countries. *Waste Management* 29, 2, 915–923.

Tsapekos, P., P. G. Kougias, S. A. Vasileiou, L. Treu, S. Campanaro, G. Lyberatos and I. Angelidaki. 2017. Bio-augmentation with hydrolytic microbes to improve the anaerobic biodegradability of lignocellulosic agricultural residues. *Bioresource Technology* 234, 350–359.

UK Department for Environment Food & Rural Affairs. 2017. Sewage sludge on farmland: code of practice for England, Wales and Northern Ireland. London: UK Government publication. Retrieved on November 30, 2017, from https://www.gov.uk/government/publications/sewage-sludge-on-farmland-code-of-practice/sewage-sludge-on-farmland-code-of-practice....

Vivekanand, V., P. Ryden, S. J. Horn, H. S. Tapp, N. Wellner, V. G. H. Eijsink and K. W. Waldron. 2012. Impact of steam explosion on biogas production from rape straw in relation to changes in chemical composition. *Bioresource Technology* 123, 608–615.

von Sperling, M. and C. A. de Lemos Chernicharo. 2005. *Biological Wastewater Treatment in Warm Climate Regions*. London: IWA Publishing.

Weiland, P. 2001. Grundlagen der Methangärung—Biologie und Substrate (Fundamentals of methane fermentation—biology and substrates). *VDI-Berichte* 1620, 19–32.

Wheatley, A. 1991. *Anaerobic Digestion: A Waste Treatment Technology*. Essex: SCI and Elsevier.

World Health Organization (WHO). 2000. Air quality guidelines for Europe, 2nd ed. European Series, No. 91, 146–147. Regional Office for Europe Copenhagen: WHO Regional Publications.

Zhang, J., R. B. Guo, Y. L. Qiu, J. T. Qiao, X. Z. Yuan, X. S. Shi and C. S. Wang. 2015a. Bio-augmentation with acetate-type fermentation bacterium *Acetobacteroides hydrogenigenes* improves methane production from corn straw. *Bioresource Technology* 179, 306–313.

Zhang, Q., M. Benoit, K. D. O. Vigier, J. Barrault, G. Jégou, M. Phillipe and F. Jérôme. 2013. Pre-treatment of microcrystalline cellulose by ultrasound: effect of particle size in heterogeneously-catalyzed hydrolysis of cellulose to glucose. *Green Chemistry* 15, 963–969.

Zhang, Y., L. Zhang and A. Li. 2015b. Enhanced anaerobic digestion of food wastes by trace metal elements supplementation and reduced metals dosage by green chelating agent [S,S]-EDDS via improving metals bioavailability. *Water Research* 84, 266–277.

Subject Index

A

Abattoir wastes (*see also* Biogas yield), 52

Absorption (*see also* Biosorption, and Wet scrubbing)

ABR (*see* Anaerobic baffled reactor)

Acetogenesis (*see* Anaerobic processes), 3–5

Acetogenic microorganisms, 3, 5
  Methanobacillus omelionskii, 5
  Clostridium, 5
  Syntrophomonas buswelii, 5
  Syntrophomonas wolfei, 5
  Syntrophomonas wolinii, 5

Acidogenesis (*see* Anaerobic processes), 3–5

Acidogenic microorganisms, 3, 5
  Bacillus, 5
  Butyrivibrio, 5
  Clostridium, 5
  Eubacterium, 5
  Desulfobacter, 5
  Desulforomonas, 5
  Desulfovibrio, 5
  Lactabaccillius, 5
  Pelobacter, 5
  Pseudomonas, 5
  Sarcina, 5
  Staphylococcus, 5
  Selenomonas, 5
  Streptococcus, 5
  Veillonella, 5

AD (*See* Anaerobic digestion)

Adsorption (*see* Dry scrubbing)

Aerated lagoons (*see* Effluent posttreatment)

Aerobic composting or digestion (*see also* Digestate), 19, 58, 83

Aerobic microorganisms, 17
  autotrophs, 17
  heterotrophs, 17

Aerobic processes (*see also* Effluent posttreatment), 16–21
  carbonaceous oxidation, 16–20, 80

effluent polishing (*see also* Effluent posttreatment), 80

endogenous respiration, 17, 83

lysis (*see* Endogenous respiration)

nitrification (or Nitrogenous oxidation), 16–20

Aerobic pond (*see* Aerobic processes, and Effluent posttreatment)

Aerobic pretreatment (*see* Pretreatment methods)

Agricultural residues (*see also* Biogas yield) (*see also* Worked examples), 52

Alcohols, 3

Alkalinity (*see also* Anaerobic treatment process control and performance indicators), 9, 26, 38

Aluminum hydroxide (or Alum) (*see* Coagulants and coagulation)

Ammonia (*see also* Health & Safety, and Nitrification), 4–22, 79
  effect of pH, 13
  toxicity in anaerobic treatment processes, 12

Ammonification, 21

Amino acids, 3

Anaerobic-Aerobic system (*see* Anaerobic digestion)

Anaerobic baffled reactor (*see* Anaerobic wastewater treatment systems)

Anaerobic covered lagoon or pond (*see also* Septic tank), 90

Anaerobic digestion (or Anaerobic treatment of solids and slurries), 41–53
  batch operation, 52
  biogas potential (*see also* BMP), 50–52
  co-digestion, 42
  combined anaerobic-aerobic systems, 47
  continuous operation, 53
  digestate, 41

Printed in the United States
by Baker & Taylor Publisher Services